AF693648

A LIFE AT FULL SPEED

A Journey of Struggle and Discovery

Charles M. Herzfeld

Potomac Institute Press

Copyright © 2014, Charles M. Herzfeld, All rights reserved.

Publisher's Cataloging-in-Publication Data

Herzfeld, Charles M.

A life at full speed : a journey of struggle and discovery / Charles M. Herzfeld. -- [Arlington, Virginia] : Potomac Institute Press, [2014]

pages ; cm.

ISBN: 978-0-9898556-3-1

Includes index.

Summary: As director of ARPA, he was the force behind the development of the ARPANET, which ultimately became the Internet we know today. Here is the story behind this phenomenal achievement, and also the story of a remarkable life. It begins with his escape from the brutality of wartime Europe in 1942, and takes him on a journey to the center of the technologies that transformed mid-20th century America: the Internet, the supercomputer, missile defense and precision warfare.--Publisher.

1. Herzfeld, Charles M. 2. Scientists--United States--Biography. 3. Computer scientists--United States--Biography. 4. United States. Advanced Research Projects Agency--History. 5. ARPANET (Computer network)--History. 6. Internet--History. 7. Cold War--Technological innovations. 8. Military research--United States. 9. Defense industries--United States--Technological innovations. 10. Technological innovations--United States. I. Title.

QA76.2.H47 H47 2014

004.092--dc23 2014

This book contains information obtained from the author's personal papers and photo collection. The author and publisher gratefully acknowledge the permission granted to reproduce copyrighted material in this book. Every effort has been made to trace copyright holders and to obtain their permission for use of this material. The author and publisher apologize for any omissions and would be grateful if notified of any corrections that should be incorporated in future reprints or editions of this book. Government-produced materials are not copyright protected (17 U.S.C. § 105). The author and publisher do not hold the copyright to works of the United States government appearing in this publication. The author and publisher cannot assume responsibility for the validity of all materials or the consequences of their use; the view and opinions expressed do not necessarily reflect those of the Potomac Institute for Policy Studies and the Potomac Institute Press.

Cover images: *Internet image map, © 2013 Barret Lyon (The Opte Project [CC-BY-2.5 (http://creativecommons.org/licenses/by/2.5)], via Wikimedia Commons),* and (superimposed) *ARPA Net 4-Node Map.*

Published, 2014 by the Potomac Institute Press

POTOMAC INSTITUTE FOR POLICY STUDIES
901 N. Stuart St, Suite 200
Arlington, VA, 22203
www.potomacinstitute.org
Telephone: 703.525.0770; Fax: 703.525.0299
Email: webmaster@potomacinstitute.org

I dedicate this to my wife Shannon, who has encouraged and helped me in every way possible, she is the wind in my sails.

TABLE OF CONTENTS

About the cover image

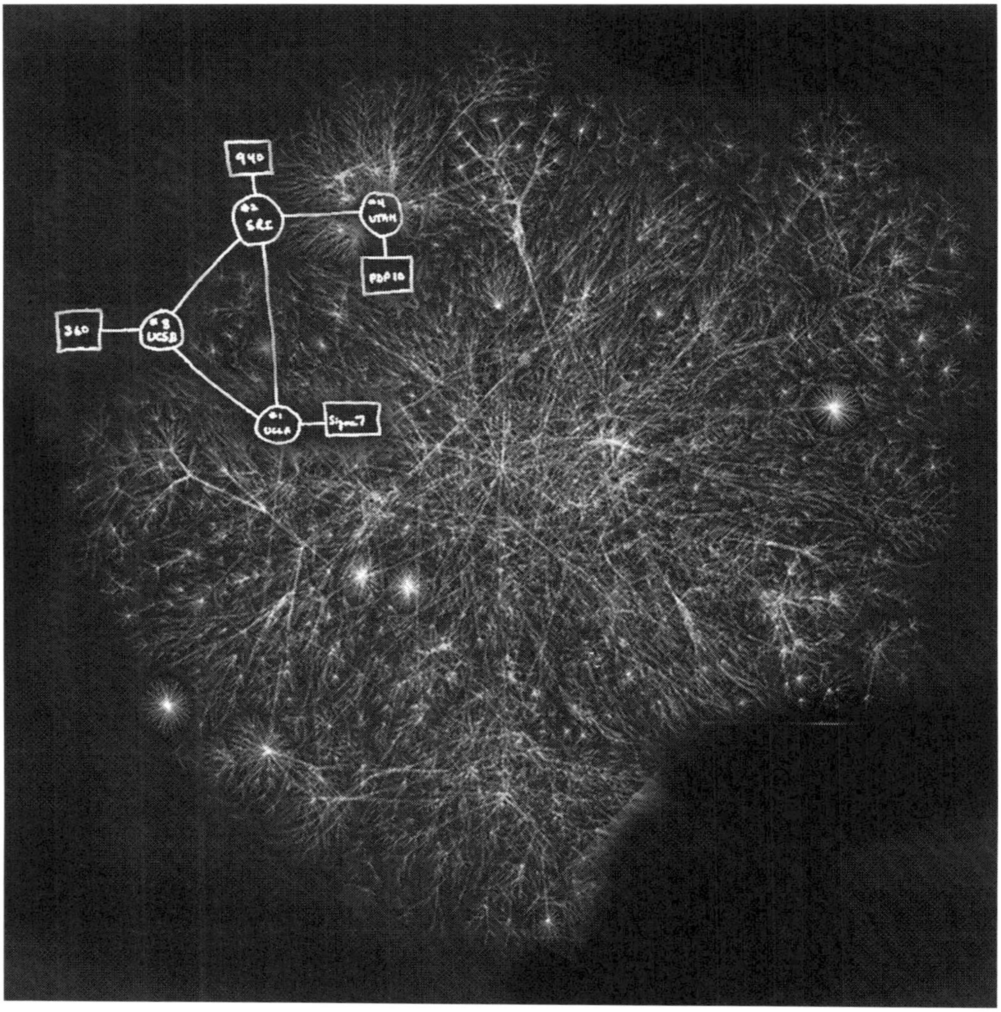

The picture on the front cover has a key thematic point. The Internet, in one of its more beautiful pictures, provides the deep background. The small diagram in the upper left corner is the original schematic of the first version of the ARPA NET that we started and built.

Internet image map, © 2013 Barret Lyon (The Opte Project [CC-BY-2.5 (http://creativecommons.org/licenses/by/2.5)], via Wikimedia Commons).

ACKNOWLEDGEMENTS

Several people deserve very special credit. First of all, my wife, Shannon, who has made my life a wonderfully happy time. This book is dedicated to her. Her encouragement and help with all of life, shielding my time for working on this book was critical, as well as her enthusiasm and understanding.

Next, my thanks to John Ketcham, who interviewed me for this book, and who took on the many hours of recording my reactions to his questions. These were very strategic in the classical sense of the word. For example, once, out of the blue he asked: "Charley, when did the Cold War start for you?" (Long and complex answer: in occupied Yugoslavia, in 1942.) Later: "How did a kid from Vienna (think snow and skis) wind up an ocean person?" (It initially reflected the interests and vacation plans of both my parents, and finally of myself.) These and many others helped structure the whole effort, and his continuing probing was critical.

Jim Giordano supported the project in many ways, and Sherry Loveless, my editor, took on the challenge and turned it all into readable prose. Many others helped in critical ways: Kathryn Schiller-Wurster arranged for the transcriptions of the dictations into editable text, a truly large challenge. Mike Swetnam, boss and friend, gave me much critically important advice on content and approaches, and earnestly encouraged me to actually do this.

Three early supporters deserve mention: John Lesko, who early on encouraged me to present my views of the start of ARPANET at a Battelle conference, Amy Friedlander, who early on encouraged me to write some very early observations, and Mariann Unterluggauer, who made my contributions to the Information Age accessible in Europe.

As I worked through the material, I became impressed how many people helped me all through my life; I have tried to acknowledge as many as possible by mentioning them individually. The usual broad admission applies, the errors are mine, the good stuff happened because of the help.

Introduction by John Ketcham

This is a book of memories, recollections covering over 80 years. Memories are tricky; some are clear, some clouded by time, some are embellished with extraneous facts and others are fragmented: many are forgotten. Memories often trigger related thoughts. They take you down unexpected paths and these paths have surprising branches, too. These are Charles' memories.

My role in this was to try to structure interview questions to elicit his memories and thoughts, to follow those various paths eventually returning to a starting point. Sometimes finding our way back took weeks. The real task was organizing these slippery tidbits of insight into their own meaningful timeline.

I first met Charles in January, 2004. I was producing a video for DARPA, interviewing a few past Directors explaining how DARPA was unique. He said, "DARPA was set up to think outside the box. It's not supposed to do too much thinking inside the box at all." Our next meeting was in January, 2006. We were back in the studio for a more in depth interview as part of ARPA/DARPA's 50th anniversary.

A few months after the anniversary banquet, Charles called me and asked if I'd be interested in helping him with his autobiography. I agreed.

We met once a week. We would come up with a topic and I would think up some questions for an interview. We tried to focus the interview topics each week, but because it was the stuff of memory, one thought often elicited other thoughts and we would often drift down unexpected and unchartered paths. This was true of both Charles and me. His responses would bring up unscripted questions which would bring up entirely new topic areas to explore. We recorded the interviews which were then transcribed.

A good portion of the material is centered in the Cold War. This occupied a healthy chunk of his professional life and a good part of my youth: I grew up with nuclear weapons hanging over my head, with Vietnam, the arms race and the eventual collapse of the Soviet Union. Here was someone with first-hand knowledge of what was actually happening. His memories would often trigger my own. Here was my opportunity to have a peek behind the curtain of the World's largest stage! (He would more often than not quickly correct my misunderstandings!)

This is more than reminiscing over the past, here are insights to a way of thinking and a way of living. We talked many times about the art of "doing" DARPA; politics and policy making are art forms. But there is another "art" offered up, too, the "art" of living a full, productive life.

The role of performing on the World's Stage is one thing, but not the only thing. The chapters about his love of the sea, from sailing to scuba diving suggest the need to balance one's life. As was typical with our methodology, once I asked, somewhat incredulously, "So how did a guy from the Austrian Alps get so interested in the ocean?" His answer sets the tone for most of what follows: "Well, let me tell you!"

This is the story of Charles Herzfeld; his childhood, his escape from the Nazis, his relocation in the United States, studying chemistry and physics at the University of Chicago (when Fermi and Teller were there!), joining ARPA, working in ITT and then again in the DoD – the breadth and depth is truly amazing. I asked him once why he seemed to take on so much. He said, "Because it was fun."

My role in all of this was to simply elicit these wonderful stories.

Figure S.1. My Mystery Island, which turned out to be the (formerly Venetian) San Giorgio.

Figure S.2. Shannon at Sounion.

STARTERS

It was on August 1, 2003, in the morning, in Vouliagmeny, a resort town in Greece, when I really started writing all this. It was a sunny, cool day. I was sitting on the balcony of our room in the Margi hotel. You could see the coast of Greece, south of Athens, the ocean, and in the far distance, what became my "mystery island," small, very mountainous, far away.

My wife, Shannon, was attending a conference in Vouliagmeny. Her meeting was held in the fancy resort hotels of Vouliagmeny (with a much better view than the Margi.) I was on my own for most of the day. I had hoped to finally get going on the "book." That day I started to write Chapter 1 about the long trip out of Europe during World War II, when my mother, our dog and I made it out. This was clearly a good place to start the book since if this trip had not succeeded, there would have been no later chapters. I had gone through many false starts writing this, and I wanted to really get going. So, I started on what I hoped would be a Christmas present for the family, a description of "our great escape," a story I had told many times, and everybody had said to me: "Write it down."

So I started, and as I wrote, the story got longer. I realized that what made this story my own was the context – what else was going on, and what I was thinking then. The telling of the events brought with it ever more detailed descriptions of context, so it became pretty long. Well, the story was done by that Christmas, and was distributed to the family in "Staples®" publication format.

I also showed this chapter to my friend and boss, Mike Swetnam, Chairman and CEO of the Potomac Institute for Policy Studies. He thought it had promise and he gave me very good advice on several key aspects, and encouraged me to go on with it. This turned out to be a critical factor encouraging me.

Where to begin? Best begin at the beginning. Born in Vienna, Austria in 1925, I had come to America late in 1942, in the middle of World War II, seventeen years old, with a spotty education, shaky English with a European accent, very old fashioned in many ways, and with no prospects at all. I wanted to become an aeronautical engineer, the war was going strong, and airplanes were important. I had already flown in several kinds of planes, and I wanted to build the best. America was a strange land for me, most of my knowledge of it came from American movies. My father had worked in America in the 1930s for a while, and wanted to return, because he thought it was "the land of the future," however he had died in a Budapest hospital in 1940. My uncle Karl (my father's older brother) and his American wife Regina were in Washington, DC, both teaching at Catholic University – he Physics, she Anthropology. And they had been instrumental in getting us – my mother, our Hungarian sheep dog and me – out of Europe and to America.

The next step was to get me educated, and get in tune with America. This seemed to take a very long time. First came Catholic University, for a rushed, three full semesters per year, course of study, for an eight-semester degree in Chemical Engineering, which I completed *cum laude*. The war was going hot and heavy, and the Army looked me over but rejected me for health reasons, as I had picked up tuberculosis (TB) in Europe. Then the war ended with the use of the A-bomb, and I went to the University of Chicago for graduate studies, because I wanted to study with the giants in science who had worked on the "bomb." I was not eager to get into the "bomb"-field as such, but was drawn by the talent and reputations of the scientific leaders of that effort. The effect Chicago had on me was near magical – I learned a huge amount, got a great and deep perspective of science, and learned how to learn – and that was the most important lesson. Also, I became dedicated to science, and research became my life goal.

Then came the Army again. World War II was over, but the Cold War had heated up unmistakably with the attack of North Korea on South Korea and the American Army there. This time the TB was gone, a major good break for me, and I was drafted. The Army trained me for infantry combat, (by that time mainly against the Chinese "volunteers" fighting against us in Korea) but then sent me to the Ballistics Research Lab in Aberdeen, Maryland, where I started doing research in the theory of interior ballistics. It was a new field for me, but I learned enough to be able to make a major contribution. It was like a weird post-doctoral fellowship. Weird, because such a fellowship is usually a continuation of work in your thesis field, but I had a complete change of scientific area. This weird type of break will recur again in other forms, it is a kind of theme of my life.

After two years in the Army, I got a job – a good one – at the Naval Research Lab in Washington. I had two moderately productive years there, and then I moved to the National Bureau of Standards (NBS) in Washington, which began a very exciting tour. NBS was then, and still is, the best lab in the US Government – it is now called NIST, the National Institute for Standards and Technology. I did some really good research, published papers, which got a lot of favorable attention, and began to steer research programs. Also, I began to teach Physics in the evening through the University of Maryland, which I enjoyed a lot. I found that the professional life of a scientist gave me a road to a good life, satisfying in most ways. I had finally gotten my act together, and I saw the way to live according to my standards and values.

Everything changed again in 1961 with a portentous phone call from "a Doctor Ruina from ARPA," asking me to come over and talk to him. He offered me a big job at ARPA (Advanced Research Projects Agency). At first I declined the offer. But that summer I went to Europe, to take some NBS Platinum resistance thermometers on a round robin calibration of our instruments comparing them to those of several European Standards Labs (and also take in some vacation travel). While in Paris, the Berlin Wall was put up suddenly, and after thinking about it some more, I accepted Ruina's offer. That was to lead to a large change in my life. But I was still in government service and in Washington.

ARPA played a unique role in my professional and personal life. Working there was the most challenging and exciting job of my life. It enabled me to do my best work, with the most important consequences. It was the first (perhaps the only) job that required all my energy, all my ability, all my determination, and made it worthwhile to engage totally. It challenged me completely. Over time it became a shining star over my life. It also cast some deep shadows. With power, responsibility and action came criticism and opposition, some helpful and well meaning, some not so. Friendships and enmities grew that have lasted.

Looking back, I am pleased and satisfied with the results we achieved, and with the recognition and credit that we received, although some of the recognition was quite slow in coming. And the failures stand out in sharper relief. Those times and events are only now being seen by others in the perspective we brought to them originally. Old differences are becoming mellowed, at least on my side.

The ARPA story is a complicated one, mostly because the challenges and opportunities were so diverse and so vast. It is hard to imagine today what was asked of us then, and what was possible then to work on usefully.

I remind my readers that I do not claim to write "history" in the sense that a professional historian would, with documentation and reference citations, but that I try to report carefully what I think about the events and my thoughts and actions then, as it was all happening. It is really my story. This report is based in part on memory (always a somewhat suspect voice), but it is buttressed with a large volume of backup from my records and notes, with many matters in the public domain, and from private sources. I take care be accurate, and when my personal stakes are involved they are briefly reported.

The main feature of that time, now barely imaginable, was that we were expected to help SOLVE the problems we were assigned (great challenges like ballistic missile defense, or detection of nuclear explosions, or helping the South Vietnamese, or starting ARPA-net), not just work on them, we were expected to "rebuild the house," not just to "rearrange the furniture." The contrast with today is remarkable.

When I left ARPA, battle fatigue and a rapidly shrinking net worth were the main drivers, and the feeling that I had served in government long enough. I wanted to work in a modern multi-national and multi-product firm, and found one: the legendary ITT of the late nineteen sixties. ITT stood originally for International Telephone and Telegraph Company, but in addition to telecom products it was also active in defense, as well as insurance, Wonder® bread, Avis® car rental and many others. There, I had a long career, starting from defense, later to include energy products, telecommunications, and finally a leadership role in corporate technical management. It turned out to be major productive change of focus and career, and I found most of it very satisfying and enjoyable.

Shortly after I went to the private sector, I also took up scuba diving, a sport that crystallized my involvement with the ocean. My under-sea life became my main sport, underwater photography a major hobby, and things nautical reconnected me smoothly with early interests in exploration. After a few years, I landed on Bonaire, in the Netherlands Antilles, a spectacular place to dive and to decompress. And while diving in many places, I returned to Bonaire occasionally. Then, in 1990, my wife, Shannon, and I bought a condo in Bonaire. We have come back many times, and it has become our second home, a haven in stress and a glory in easy times, and we go there still.

After ITT, came consulting for government, and for private firms. This turned out to be for me a great machine for growth and learning. I became part of a large network of fellow experts, with wide interests and deep knowledge, and all that was great fun. I traveled extensively and comfortably.

But in the late 1980s things changed again. The Cold War, a life fixture for me, began to wind down, and I got intrigued by participating in managing this epochal event. George HW Bush had become President, many of my friends went into the administration, and I decided to try for the top technical position in the Pentagon, Director of Defense Research and Engineering. This transition was slow, and the results were complex. The war with Iraq came early in my tenure and put a spin on all things to follow. Some of my bosses were more concerned with orderly appearance than real solutions, and I was mightily frustrated. I could not get done the things I thought should have gotten done; but I had been stuck before. A ten-month tour in the White House, working for Alan Bromley, the President's Science Advisor opened up, and I got some things done there that were really important. Then it became time again to leave the government, and I became a consultant with interesting assignments with Los Alamos National Lab and Sandia National Labs. Now, finally, I was working with the real "bomb" people.

And eventually the realization came that it was time to think about the things that happened, and how they led me into the life I had. It was a time to reflect, and start the winding down process. This book is the outcome of this process. As I look over a long and very busy life, several things stand out: At every stage I had a lot to learn, and luckily had excellent teachers, mentors and friends at all parts of the process. I did a lot of things and worked with a great variety of wonderful people, to whom I owe greatly. I have tried to mention every person who helped and who had a major impact on me. I do this as a "super-acknowledgement," – thanks to all.

A final admission: all mistakes are mine, and all the good had lots of help.

Charles Herzfeld
Silver Spring, Maryland
January, 2014

CHAPTER ONE

The Escape from Wartime Europe

Begun in Hotel Margi, Vouliagmeny, Greece, August 1, 2003.

It is October 11, 1942. I am sitting on the steps of an apartment building in Baltimore, with my mother and our dog. We have just arrived from an arduous and dangerous journey out of wartime Europe, after three years of war. Getting to the apartment had been a real challenge: strange country, strange people, our limited British (not American) English, and strange money. At my uncle's suggestion, we had taken a cab to his friend's apartment in Baltimore. The apartment was dark and locked, so we sat down on the stairs in the building and waited for my Uncle Karl and his wife Regina to arrive. They were our lifeline, the ones pulling the other end of the long string to which we had clung for over two years.

We are cold and hungry, and Betyar, our dog, is softly whimpering. He is aching for food, water, and rest. And so are we. But for now, there is nothing to do but watch and wait – something to which we have grown all too accustomed since the war began.

The Beginning: Vienna, 1938

Our journey began in March 1938, when the Germans marched into Austria and annexed the country. To my amazement, many

Figure 1.1. The front of the house in Vienna, taken in 1993, unchanged from 1938, my room was on 2nd floor, front right corner. CH personal collection.

Austrians welcomed the Germans, something I did not understand until much later.

In August 1938, I moved to Budapest with my parents and Betyar, our Hungarian sheepdog. My father's employer, the Standard Oil Company of New York, had transferred him to Budapest. We got out just in time – had my father not been transferred, the Germans would surely have grabbed him to get at his knowledge of future fuel requirements of Europe's air forces.

We were able to establish a fairly normal life in Budapest, we lived in a nice house, and I went to a nice school.

But tragedy struck our family in the winter of 1940. My father entered the hospital for a minor operation to fix an old injury he had suffered on the Western Front, during World War I. The surgery was successful. However, as he was recuperating, a small outbreak of meningitis spread through Budapest hospitals. Antibiotics were unknown at that time, and there was no cure. My father died a week later, on January 30, 1940. We moved to a smaller house, and I continued in my school.

Suddenly, the question loomed before us: stay or go? At that time, it appeared we could stay safely in Hungary for a few years. But by the middle of 1940, Hitler controlled much of Europe, and had begun to turn the captured nations (Poland, France, Holland and Belgium, Norway and Denmark) into German colonies. Hitler began to impose the Nazi terror regime on these occupied countries, and he had clearly announced his ambitions to spread Nazi rule and suppression across Europe. There were many credible reports of persecution and outright elimination of all opposition in concentration camps, and threats of more to come. It was only a matter of time before Hungary would fall victim. Some of my fathers' friends had spent time in Dachau, the Nazi's first concentration camp, and they explained the mistreatment they had suffered, such as being kept outside in the extreme cold of winter, with their bare feet in buckets of water and ice.

Figure 1.2. My father, mother, and Betyar in Vienna, in 1936. Photo by CH.

I vividly remember one discussion with my mother, during which she took me aside and said, "Look, we can leave Budapest, or we can go to Germany, and

Figure 1.3. Winter 1940: My mother, Betyar, and me. My father had just died. CH personal collection.

Figure 1.4. The Coronation Church, where my father and I often went to Mass. Photo by CH.

Above: Figure 1.5. My class in the German School in Budapest (I was ill the day Figure was taken). In the center is our teacher, Professor Billmayr, killed on the Russian front in 1944. Notable here is my friend Norbert Pohl, with whom I kept in touch over the years, and George Eissler (who later escaped to Argentina with his family). We three were great Navy buffs, and followed the war action day by day, avidly. Also in the picture is Eva, a girl that I was very fond of, but as was the custom in those days, we hardly ever spoke. The boy on the left, in the front row, is Andreas Chrambach. His entire family died in a concentration camp, but he survived, came to the United States, and became a famous biochemist. He worked at the NIH until his death in 2006. CH personal collection.

Above: Figure 1.6. The author in 1941, at age 16. CH personal collection.

you will be safe. My Onkel Adul and Tante Jettl will take care of us. What do you want to do?" It was a wonderful thing for her to let me have a big vote in the matter. We talked for a few hours in low tones (at that time we were living in a pension on Molnar Utca, and privacy was hard to come by.) The pension was an old, large and elegant apartment that had been converted into perhaps ten bedrooms, a large living room and dining room. The décor was from around 1870. The people living there were an unusual bunch, but more on this later.

The facts on the ground were clear to us: Hitler was going to try to rule all of Europe, or burn it if he failed. I was only sixteen years old. If we stayed where we were, it seemed the next stop for me would be the German army, and fighting for Hitler and his monstrous regime was not my idea of a great future. It was obvious to both my mother and me that we should do all we could to get out – and the quicker the better.

So we began to plan our departure, an effort that absorbed all our attention until we arrived in America in 1942. As we left Budapest, our decision looked even wiser. The Germans had just invaded the Soviet Union and were defeating the Soviet Army in huge battles. The largest land war in history was starting. Getting out was the right answer.

My uncle Karl and my mother had begun discussing a move right after my father died, but without a clear idea of where we should go. There were two obvious requirements: We had to go somewhere that was safe from the war and where living was cheap. We had no money at all, and were depending on Karl to support us with just his professorship at Catholic University. (Much later, we learned that Karl was also supporting his stepmother Maria, and her son, his half-brother Herbert.) We also hoped that our destination would have a good, inexpensive university nearby.

Karl and my mother had no choice but to discuss our options at a slow pace. Air mail was costly and took about two weeks. Long-distance telephone calls were exorbitant, had to be "booked ahead of time" at the City Central Post Office and were known to be monitored closely by the German Gestapo. In the spring of 1940, Karl suggested we go to France, to Avignon, which had a good university and where living would not be prohibitively expensive. But not long after he had advised

Figure 1.7. My mother with Betyar, left, and Bogar, his spouse, outside our second house in Budapest. My room is on second floor, on the right. Photo by CH.

Figure 1.8. The dining room of our second house in Budapest. Photo by CH.

Figure 1.9. The living room. Photo by CH.

Figure 1.10. The sun room. Photo by CH.

this, the Germans overran France. It suddenly became crystal clear that unless we escaped soon we would not get out.

No matter where we fled, we knew our lives would change dramatically. During our years in Budapest, we had enjoyed a pleasant, safe life. Food was cheap and very good, the people were delightful, and cultural opportunities abounded. When we first arrived in 1938, we lived in modern house and I attended an excellent school, the so-called "German School." It accepted refugee kids like me, including many from Vienna and from Prague. The teachers were excellent. Instruction was in German, but we also studied the Hungarian language and Hungarian literature and history. This was taught in addition to the standard topics: mathematics, history, geography, German literature, English, Latin, and French, which all were taught in German. This was all secondary to my biggest excitement, though: the classes were co-ed! This was my first regular contact with girls, and produced many surprises.

After France fell in 1940, life in Hungary grew much more tense. My teachers began to hew more closely to the Nazi political line.

When money became even tighter, we gave up this house. The furniture was stored in a storage facility near the Budapest railway station. However, all our belongings were destroyed in an Allied air raid in 1944.

We moved to the pension already mentioned, where we had two small rooms and a bathroom. The people living in the pension were an odd bunch. I remember two in particular: One was a retired Hungarian landowner, (I don't know why he lived there, but he must have lost his land) who enjoyed eating and talked incessantly about hunting. The other was a Canadian woman, who loudly and aggressively praised Hitler, Germany, and the Nazi regime, while criticizing the British, in a very "British" English. She was particularly vicious in her contempt of Winston Churchill. When there was news of German bombings of British towns, she expressed hope that there would be lots of casualties. It was disturbing to say the least, but it was also dangerous. It was clear to us that she would have handed anyone who was not enthusiastic about Hitler to the secret police in a heartbeat. She was apparently also a real German agent, and later did propaganda broadcasts to England for the Germans. After the war, she was tried for treason, found guilty and executed by the Canadian government. And she was a constant reminder that it was crucial we conceal our reasons for leaving Hungary from every living soul.

Meanwhile, plans for our trip continued in secrecy. Once France ceased to be a viable option, Karl and my mother pondered the possibility of our going to Brazil. It was inexpensive, politically safe, and had an excellent university in Sao Paolo. However, it was extremely difficult to obtain a Brazilian immigration visa. Finally, Karl decided that the only workable option was for us to go directly to the United States.

Once we had settled on our destination, the next critical challenge was to figure out how to get there – which led us to the problem of documentation. We had no valid passports from our homeland; Austria was no more. Requesting a new passport (which would have to be a German one) would alert the Germans that I was nearly of draft age. They would never have let me go anywhere but Germany.

Solving this problem was urgent, and here my mother's creativity and drive really shone. She proposed a daring solution: not one, but *two* passports, one neutral (Hungarian) and one Allied (Polish). It was both difficult and dangerous: if both were found on us, we were likely to be shot, or sent to a concentration camp. But we did not see any other options.

The Polish passport came first, thanks to the Polish military attaché in Budapest. His name was Kasimir Zdiarski, and he and his family were friends of my parents.

After France had fallen, the Polish government, working from exile in London, moved the Budapest Embassy to Cairo. Before Zdiarski left, he issued a valid Polish passport to us, with all supporting documentation. This became the document into which the American immigration visa was eventually put. (I still have that marvelous, life-saving item!)

But that still left the question: What document could we use for the trip from Budapest to Portugal? Portuguese ships were among the few neutral ones at sea, and the port at Lisbon was our best chance.

Again, my mother devised an implausible solution (remember, none of us had experience with false or fake passports): She "bought" a Hungarian passport from a fellow she knew, who was working in the

Hungarian Interior Ministry, and who was always deeply in debt (because of too many girl friends and too many expenses). He gave her the Hungarian passport in exchange for a magnificent diamond ring my mother had. This was no fake passport – the paper, the printing, and the numbers all were real. But there was zero documentation backing it up. Any careful check would have revealed that it was nothing more than a piece of paper. Still, it was a risk we had to take. (For reasons that will become clear, we did not keep this document.)

The plan was to leave Hungary and travel through Yugoslavia to Italy. From there, we would go through France and Spain to Portugal, and then sail from Lisbon to America. We had to have a transit visa from each country we entered. And in order to issue a transit visa, each country had to see such a transit visa from the next country in the chain. Thus, Italy would be happy to give us a transit visa – provided we could show that we could either go back to Hungary or forward to France, and so on for every transit.

The American visa came through the hard work of Karl, helped by some VIPs at Catholic University. But we could show neither the American visa nor the Polish passport to anyone without exposing our end goal, and thereby getting arrested. The problem was solved by the very kind Portuguese consul in Budapest, who had seen our American immigration visa in the *Polish* passport, and gave us a Portuguese visa in the *Hungarian* passport. After seeing the Portuguese visa, Spain then issued us a transit visa, working backward in this manner.

The next challenge was France. After the collapse of France in 1940, the northern and western parts of the country were occupied by the Germans and were under strict Nazi control. The central and southern parts became "Vichy France," run by a new French Nazi government (located in the old French spa of Vichy) which was craven in its support of German policies and war aims. Getting a Vichy visa looked impossible. Karl worked diligently to obtain one, but even his contacts in the American Embassy in Vichy and the Vatican had no luck. We got so desperate that we even contemplated going through the Soviet Union. This might have been feasible before June 1941, when Germany attacked the Soviet Union. Two of my Dutch school friends from Budapest, and their families, actually made it out of Hungary that way. (But they were unlucky – they

Figure 1.11. Mr Zdiarski, his wife and son, and my mother. Photo by CH.

got to their destination, the Dutch East Indies, now Indonesia, just a little time before the Japanese invaded it. I never found out what happened to them.)

But then I read in the newspaper that the Vichy government had appointed Jean Borotra as minister for sports. Borotra had been a world-class tennis champion in the 1920s, and then had gone into business running a large chain of gasoline stations in France. I had heard my father talk about him – he knew him from his work at the Standard Oil Company. I reminded my mother of this connection, wondering if he could help us get a Vichy transit visa. This was indeed a very long shot. She wrote to him, and after an excruciating long wait he indeed arranged to have a visa sent. With this in hand, Italy was ready to issue our next transit visa. It was now early 1941.

But then came another roadblock. All throughout the spring of 1941, German troops had moved through Budapest on trains and trucks going south, their destination unknown to us. In April 1941, Hitler invaded Yugoslavia and Greece, and quickly overcome their resistance. German arms seemed invincible. They took over Bulgaria and Romania as occupied allies. All this was triggered by the war in North Africa, between the British and Italians (which was going poorly for the Italians) and the war between Italy and Greece (which Mussolini began in a fit of hubris in late 1939, and which also was going badly for him). So the trap had sprung on us. There was no way out through any country that was not under

Figure 1.12 and 1.13. The Polish passport, inside pages and outside cover, on a chart of Chesapeake Bay, with Baltimore. Photos by CH.

German control. We pondered what to do next, and sometimes when it just seemed too hard, we thought about giving up altogether. But the war in the Balkans reminded anyone who still needed reminding of the terrible repression under German occupation.

Meanwhile, the situation in Yugoslavia had settled somewhat. The Germans and Italians had occupied all of it quickly, after a stiff – but brief and ineffective – resistance. They divided the country into spheres of influence: Italy occupied the land along the Adriatic coast down to Albania and, with the Germans, created a Nazi puppet state in Croatia. This state shared a border with Hungary and Italy. Germany occupied Serbia and the rest of the Balkans. So our escape plan was still viable, if we could go through Croatia, where there was constant and bitter guerrilla fighting. This turned out to be possible with the Hungarian passport.

But another geopolitical earthquake shook the world on June 21, 1941. Hitler attacked the Soviet Union in a great surprise move. (A surprise, because Hitler and Stalin had signed a secret pact in August 1939, before they attacked and overran Poland. No one expected this sudden switch, least of all Stalin, who was taken totally by surprise.) And so a truly major war began, very near to us. After this, most governments wanted to re-evaluate things like transit visas. All this took more time, but we were able to leave Budapest on August 20, 1941. The plan was to go first to Zagreb and then to Crikvenica, on the Croatian coast, and stay there until the entire trip was lined up – something that would take a long while, because of delays with our visas.

We arrived in Crikvenica on August 26, and stayed there until early March 1942. It felt like a million years. We felt a little safer, but not much. Italian police and military patrolled constantly. A 10:00 p.m. curfew was in place. The sound of gunfire from attacks on the Italians was a nightly occurrence. Food rationing was severe, and would get drastic. Rumors abounded that the Gestapo was working through the area to find refugees. Some people disappeared without explanation.

It was here that the decision to bring Betyar proved smart. I vividly remember an evening when I was on my own, my mother having returned to Budapest briefly to deal with some visa matters. Coming home from the bus station well after the curfew, I spotted a large patrol of Carabinieri, the Italian national police. I decided to brazen it out and walk as loudly as possible, crunching boldly through the snow with the large Betyar by my side. When I met the patrol, I just waved – and they waved me on. They had gotten to know the dog.

In the meantime, other terrible and frightening news was coming to us. Friends in Budapest who had been arrested and deported wrote us postcards as they were moving, saying they were in trains going east. They did not know where they were going or what was going to happen. And then the postcards stopped. We began to understand that something hidden was going on, something truly terrible, a mass extermination program. At the same time the Croatian press printed stories bragging that thousands of "Jews, Communists, Freemasons and Homosexuals" had been killed in the camp in Jasenovac, a camp that was not far from where we waited. All these were hints of the start of the Holocaust.

Crikvenica, August 1941 – March 1942

But for the moment, we were safe. We basked in the beautiful weather and spent lots of time on the beach.

The town of Crikvenica was a tourist attraction of sorts, very Mediterranean in style, comfortable, with many good restaurants.

The main problem was that food was getting very scarce, with tight rationing. And everywhere there were troops and Carabinieri.

We hoped to move on quickly, but it was not to be. Visa problems hung us up for months on end. We had missed the ship from Portugal to the United States, so we needed to be re-booked on the next ship out of Portugal, which triggered review of all the visas for the rest of the trip.

Life in Crikvenica got harder and harder. The weather grew cold – one night the water in Betyar's water bowl froze over, and I had to break the ice with my fist so that he could drink. Food became scarce. Restaurants began to serve horse meat in sausages. They tasted funny, but at least it was real meat. There was no fruit, few vegetables, and even the cornbread was stale.

But there were blessings to be found. That Christmas my mother was able to give me an egg. I had begun to dream of eggs, which we had not had for months. I could not have asked for a more a wonderful present! Also, we became friendly with an Italian officer staying at our hotel, a Captain Alberto, whose last name I no longer recall. He was in charge of all the army transport up and down the coast. He was a pleasant man, without any enthusiasm for the war, because he had served in the war against Ethiopia, then in the Spanish Civil War, and was now stuck in Croatia.

But whenever he could, he arranged food for the dog, and also excellent Army bread from his soldiers' mess hall. This was a life-saver. Once, they gave me a large piece of parmesan cheese, again "for the dog," Little did we know, Betyar wound up eating it despite my best efforts to safeguard it. I had put it on top of a large wardrobe and went out. When I returned, the dog was quite ill. Somehow he had climbed up on top of the wardrobe, and gotten a good bit of the cheese. Later, I ate gladly what was left.

Outside our small world, the war continued. The Germans were destroying the Soviet Army, taking several million prisoners, and advancing toward Moscow. They showed pictures, taken by advanced patrols, of Moscow in the distance, with the towers of the Kremlin in the misty distance. Yet they did not succeed in taking it. This was the first major setback for the German military, and a sign of hope.

Figure 1.14. The beach at Crikvenica, full of rocks. Photo by CH.

Figure 1.15. Occasional sailing ship. Photo by CH.

Figure 1.16. Harbor of Crikvenica: My mother, Betyar, and Felix Schorr, a friend. Photo by CH.

But that hope was short-lived. In December, the Germans reported that the Japanese had sunk the American fleet in Pearl Harbor in a surprise attack. Our first reaction was that it couldn't possibly be true, that it must be propaganda. But it was true. And more disasters followed: Japanese landings in Malaysia, Indonesia, and the Philippines. There seemed to be no stopping them. All this encouraged the local Nazis and their supporters. Life became more dangerous and frightening. Would we ever get out? And if we did, would the war be already lost?

We took heart when we began to hear news of the American rearmament. Roosevelt was planning to build tens of thousands of aircraft, tanks, and many ships. Meanwhile, the German propaganda machine turned out vast amounts of newsreels (then the most powerful and visually gripping material, all expertly edited), and also a new monthly magazine, *Signal*, the first such *ever* published with *all* pictures in color. The Nazis had a lot of good news from their point of view, they had an easy story to tell, and they told it terrifyingly well.

I was, however, able to hear largely straight news thanks to BBC broadcasts, which I listened to on our radio. The BBC seemed objective and tersely factual, with a minimum of propaganda overlay.

The BBC became both my lifeline and my sanity check. But listening to "enemy radio" was dangerous business. We were in occupied territory in the middle of a war. Here, if the Germans found you listening to the BBC, they shot you on the spot. Rumors even began to circulate that they could detect which station a radio receiver was tuned into, from the street! (This seemed implausible, but apparently was true for the primitive radios available then.) It was risky, but it was essential to learn any and every scrap of truth. I learned to be careful, quickly turning the radio off and tuning it to a different station when I heard noise in the hallway.

The radio was also the path to music, for which I developed a passion. During the early part of the war, opera was performed in most major cities all over Europe, and was broadcast live in the evening, so there was a flood of great opera available. Verdi became my favorite. By 1941, in Crikvenica, opera broadcasts had largely disappeared, but Italian radio broadcasted a great deal of music. And a special treat of live music came from the Italian soldiers, who congregated near the harbor in the evenings, and sang melancholy Italian songs about life, love, and longing for home. These songs greatly moved me, and let me see these Italian kids not as monsters bent on destroying the world, but as victims. Like everyone else, they wanted to go home.

Various Resistances

Over time, we got to know some members of the resistance. I clearly remember one of them, a man named Ivan who led local the Communist resistance in the area. He was executed by the Italians – along with two others, on Christmas Day, 1941 – "to send a message."

That same winter I met an engineer who was in charge of building a hydroelectric plant in the hills of central Croatia. One day, he invited me out to visit the construction site. It was a large place, with tunnels, large halls for the machinery, and the beginnings of a dam for the river, whose energy would drive turbines that would generate electricity for the whole area. I learned a great deal about the plant – everything from where the turbines were being built (in Switzerland, by Brown Boveri), to the expected power output, to the planned

Figure 1.17. Mother and I and two chickens in danger of being eaten, in Zagreb 1942. CH personal collection.

date for completion. He tried to recruit me to work at the plant as an engineer, even though I was still in high school. He was an example of passive resistance, no enthusiasm for the war or either side. I planned to describe this plant to the Allies, if ever I got out, and did so. Much later, I found out that the Brown Boveri machinery never made it to Croatia. The British bombed and destroyed the German factory that made the specialty railroad cars needed to transport the turbines.

One Step Back, Two Steps Forward

Finally, it was time to move on. We received our last visa early in March 1942. After what seemed like years, we were finally on our way, first back to Zagreb, in Croatia, then via train to Trieste and on through Italy.

Figure 1.18. Mother and I in Zagreb, 1942. CH personal collection.

Italy, March 1942

Overnight in Trieste, I went to a movie, which I remember vividly: *They Shall Have Music,* with Yasha Heifetz. It was beautiful music – Mendelssohn, with German subtitles – and was nothing short of surreal under the circumstances: Music written by a so-called Jew, performed by a so-called Jew, in an American movie, all in German-occupied Europe.

We moved on to Venice by train, then to Milan. We spent several days there, dealing with banks and visas. My mother arranged for us to attend a performance of Gluck's *Orfeo ed Euridice* at La Scala Opera, an exciting and memorable occasion. I also got my first eye glasses.

There were other memorable moments in Milan, but of a much more serious nature. For one, we weathered a real air raid by British bombers. We heard the bombs explode and the sounds of the anti-aircraft artillery from an air raid shelter near the hotel.

Another heart-stopping moment occurred when my mother came back from the police with the Italian exit visa. It turned out that it was valid only to leave Italy via the border town Domodossola, at the Swiss-Italian border. But we needed to go to Modane at the French-Italian border. Crossing the wrong border would have meant doom. The Swiss would not have permitted us to enter, and the Italians would not have let us back in, either. Fortunately, I happened to check the place on an atlas (an early lesson to check facts often), and my mother was able to return to the police and get the visa changed. And on we went.

On our way out of Milan, Betyar made another stellar contribution. As we were walking along the train station toward our railroad car, we were approached by two men in civilian suits and leather over coats, looking suspiciously like Gestapo. Obviously, we wanted to avoid any such encounter. My mother tensed up, thus alerting the dog, who went into a defensive crouch, bared his teeth, and growled. The strangers decided to pass us by. One lifted his hat in greeting, and they were gone.

We stopped overnight in Turin, where our ration coupons got us watery minestrone, a little rice, and bread that was so poor it was rolled in sawdust to keep it from falling apart. But the real excitement began on the next day. We went by train from Turin to the French border. When the train stopped at Vercelli, west of Turin, my mother took the dog for a quick walk, and they missed the train! Heart in my throat, I decided to get off at the next stop, with all our belongings, to wait for them. Happily, they were on the next train and we were reunited.

Through Vichy France to Spain

Soon thereafter we were in Vichy France. The French border guards were nasty, and did a close body search of both of us. They missed, fortunately, materials we had hidden, addresses and codes that could help the British establish contact with the Yugoslav resistance, as well as the second passport. We got through, but I remember it as absolutely the most threatening part of the trip.

I also remember clearly that we had a beautiful journey through Southern France. The weather grew warm and the land was full of blooming almond trees, with the snowy tops of the Pyrenees in the background. This beauty helped make bearable the poor food and oppressive atmosphere. We hoped Spain would be better. We hoped in vain.

Generalissimo Franco had decided to cooperate fully with the Germans, and sent everyone without proper documents back to the German police waiting on the French side of the border. On the train from Turin, we befriended a young man who was stopped by the Spanish to be sent back to France. He killed himself on the Spanish side of the border rather than go back to German control. We made it across into Spain, but our luggage remained on the Italian border, inside France. So we had to wait in Port Bou, just inside Spain, for almost a week before our trunks came through. When we had everything together, there was yet another delay: A rainstorm had washed out the railway bridge on the way to Barcelona, and we had to wait another several days before it was fixed. We were much too close to Vichy and the Germans for comfort.

We finally made it to Madrid, a beautiful city that now looked like a war zone. Soldiers with rifles stood guard at each major intersection. A tight curfew was imposed and there was little food. We managed to go to the Prado, the great museum, but most of the art had been moved to shelters, and the building was covered with mountains of sand bags. After a few days, we boarded the train to Lisbon with relief.

Portugal and Freedom

At the Portuguese border all went well. Our papers were in order and we made it over the border. Once we got there, everyone was removed from the train without explanation. We stood beside it for several hours. We learned later that this was a check by the British, who had an agreement with Portugal to search the train for papers, documents, weapons, and such. When we re-boarded the train, all our things were there and seemed undisturbed. Clearly the search teams were very skilled.

The highlight of that part of the journey was the dining car, where we had our first good, large meal in two years. I remember every detail – the soup, the roast veal, the fresh vegetables, the fruit and cheese for dessert. Seconds were offered! It was like paradise.

We arrived in Lisbon, on March 30, 1942. We went to our hotel, unpacked, and then went for a walk. We found many open-air markets that sold fresh fruit cheaply. We bought two big fresh pineapples. We had never seen fresh ones before, only canned. We took them to the hotel room and ate both in one sitting. And shortly after that we were terribly sick to our stomachs, because we were not used to large quantities of fresh tropical food.

We soon moved to a cheaper hotel where we would stay for several months. We enjoyed the city with its fabulous parks, the old buildings, and the fantastic pastries, which were even then, in the middle of the war, absolutely, fabulously good.

The first order of business was to rearrange our ship passage. Portuguese ships were the only neutral craft sailing, and had a fair chance of not getting sunk. The next ship to the United States left in September. There was nothing to do but to wait.

So we explored Lisbon, which to this day is one of my favorite cities. We met a few people from Vienna, from the Standard Oil Company, all on their way to the United States. We reconnected with the Eisslers, our friends from Budapest, who were passing through Lisbon on their way to Argentina. There were some young people, about my age, and

Left: Figure 1.19. Lisbon, author and Betyar, CH personal collection.

Below: Figure 1.20. Mother and Betyar at the ocean near Lisbon. Photo by CH.

we went to the beach and the parks, and had parties where we danced and played games – rather innocent, by today's standards.

Starting to Fight Back

We also reunited with Colonel Zdiarski, who was then stationed at the Polish Embassy in Lisbon. Zdiarski arranged for my mother and me to meet with British Embassy officials, who were debriefing refugees from central Europe, and they were interested in every detail of our trip.

They also asked me to go to an exhibit of German technical books because I would be unknown to the German security guards. I brought back a variety of brochures and reports, and even got my picture in the paper, standing a little behind the President of Portugal, who opened the exhibit.

Figure 1. 21. At the German exhibit. CH (standing back row, 2nd from left) standing a little behind the President of Portugal, who opened the exhibit. Source: Vida Mundial Ilustrada, Vol 1 (48) 16 April, 1941.

We were eager to assist in any way we could. I showed the British officials one of my German books about military airplanes, which I had brought out. It had several pictures of new German military aircraft engines that were unknown to the British, and they photographed those pages. They were also interested in the Croatian power station, which I described earlier. They wanted to hear about the guns and tanks we saw at the train station in Milan, ready to be shipped to Africa. And most importantly, we gave them the way to arrange meetings with the contacts in Zagreb we had arranged. They kept the Hungarian passport, all the ration cards, and our bus and tram tickets – any piece of official paper from "the inside." And they warned us that German spies were plentiful, and would likely try to make contact with us.

After all these, Mr. Ridley Prentice, the senior embassy official, with whom I had worked on the debriefings, invited me to the Royal British Club, where I had my first whiskey and soda. I have had many a drink since, but I will never forget that drink.

Not Quite Ready to Go

After a few weeks in Lisbon, we moved north to a town called Caldas da Rainha, (the Baths of the Queen) near the ocean.

The hotel there was large, half-empty, and quite comfortable. The weather was fine, the food was good, and there was even a movie theater that showed almost current American and British films. And after a while we began to notice groups of British men – clearly military – who would march off to do what looked like physical exercise. Every few weeks one group disappeared, and another appeared. Later, we learned these were British sailors and airmen, who had been rescued at sea and brought to Portugal for recuperation. Portugal was technically neutral, but leaned heavily toward the Allies.

And I did run into a German spy in Caldas. I had taken up tennis, there were excellent courts in the town, and one of my frequent playing partners was a woman in her thirties with close cropped grey hair – a refugee, so it seemed. However, after a while we got word from our British friends in Lisbon suggesting that I stay away from her: "You know what we mean..." and I avoided her.

In June, I began to attend St. Julian's, a British public school, in Carcavelos. Before I got to that real school, I had been reading voraciously whatever I could find. In this strange menu some books still stand out: Especially an old book"By Auto Car from Peking to Paris", from ca. 1912, and several books by Peter Felming about travel in Central Asia,

Figure 1.22. St Julian's School entrance, in 1969, just as it looked in 1942. Photo by CH.

ca. 1934, full of fascinating stories and pictures. (I did not know at the time that he had been in the British secret service. He was the older brother of Ian Fleming, the author of the 007-books, who also was in the British secret service.) But now formal education resumed.

About half of the students at St. Julian's were refugees, and the other half were Portuguese or children of British expatriates. The school was good, though it took me some time to adjust to the informality of instruction and of relations with the teachers. My official academic goal was to prepare for the matriculation exams for admission to Oxford or Cambridge. My actual goal was to cram in as much in as possible, I was starved for formal instruction.

The school's official religion was Anglican. We had a prayer meeting every morning from the Common Book of prayers, and then we sang hymns. Far away, the second Battle of Britain was underway, and Allied ships were being sunk in the Atlantic every day. In our school, we sang: "Oh, hear us as we cry to thee/for those in peril on the sea!"

and "Oh, hear us as we lift our prayer/for those in peril in the air!" I still get weepy when I hear these hymns. After the Anglican service, a Catholic teacher, Miss Lacey, took us to another room for a few Catholic prayers.

The war continued. Every day we listened to the BBC, which we could do now in safety. We heard about the British disasters in North Africa, and the Germans' plan to help Spain recover Gibraltar, at the entrance to the Mediterranean. Rumors abounded, mostly about whether or not Generalissimo Franco would get into the war on the German side. In Portugal we expected to be involved any day. Concern grew intense, and one day the teachers took us students to the beach and showed us the particular ship anchored in the river estuary where we would be taken to, if we were evacuated. It was remarkable to me that the British were planning to evacuate not just British citizens but all refugees, even during the desperate summer of 1942.

The Voyage Begins

But the summer passed in safety, and suddenly it was early September. Our ship, the *Serpa Pinto* (Painted Serpent, the name of the Maya god-hero) was to sail on the eighth of September.

As usual, Karl somehow solved a few last-minute crises regarding our visas and money for the trip. The voyage was complicated: The ship was to go from Lisbon to Casablanca, and from there to Bermuda, for British inspection, and then on to Havana, Cuba. We would go next to Vera Cruz, Mexico, and then, finally, to Baltimore.

I will always remember the splendid view of Portugal as our ship departed Lisbon. We passed the great Belém Tower, and then the beaches of Estoril, where we had gone swimming. By now we had come to love Portugal, and were surprised to find ourselves sad to leave.

Suddenly we were at sea. I spent most of my waking hours on deck, exploring the ship, saw its huge antiquated (pre-World War I) engines, and I was even allowed to come on the bridge. On the first stormy night, I still remember standing in the long corridor below deck. With every big wave, the corridor would bend and twist just as the ship did, and the ship would shake and groan as it fought its way through the heavy sea.

Figure 1.23. Our ship Serpa Pinto*, as she looked in 1942. CH personal collection.*

And a few days later we arrived in Casablanca, Morocco. Casablanca was an important French naval base. Like most of French Africa, it was Vichy-controlled territory, and therefore pro-German. We worried, of course, about potential run-ins with the French secret police. But as it turned out, our experience in Casablanca was short on danger and long on drama.

Casablanca

We entered the harbor in the evening.

The harbor was blacked out, as was the city. The ship's navigation lights were illuminated; everything else was dark. We anchored in the outer harbor, near Position F on the map in Figure 1.24. A number of French cargo ships were anchored around us.

We had been told that the outer harbor hand been mined, and not to stray from our position. We were also told that there were guns in place on the breakwater, and that the French Navy was on maximum alert – hence the blackout and radio silence.

During the night the wind picked up, and after a while it was blowing hard toward the northwest. Soon it became apparent that our anchors were not holding. It must have been about midnight and everyone went up on deck to see what was going on. We could hear the crunch as the anchors slipped (a sickening sound and feeling), and we could feel and see the ship slowly moving. We were drifting toward the breakwater.

The wind grew so strong that our engines, going full blast, could not keep us from drifting. The captain tried to raise the French harbormaster's radio, without success. No help came, nor even an acknowledgement of our predicament. Finally, in desperation, the captain lit up the whole ship, and sounded all the sirens – quite a spectacle in this blacked-out harbor.

This went on for some two hours, until finally our ship bumped into the breakwater, stern first, approximately at position G. It was not a great smashing collision – more a nudge – but you could hear the grinding of metal as the propeller hit the rocks. Finally the harbormaster woke up, raised us on radio, and put powerful searchlights on us.

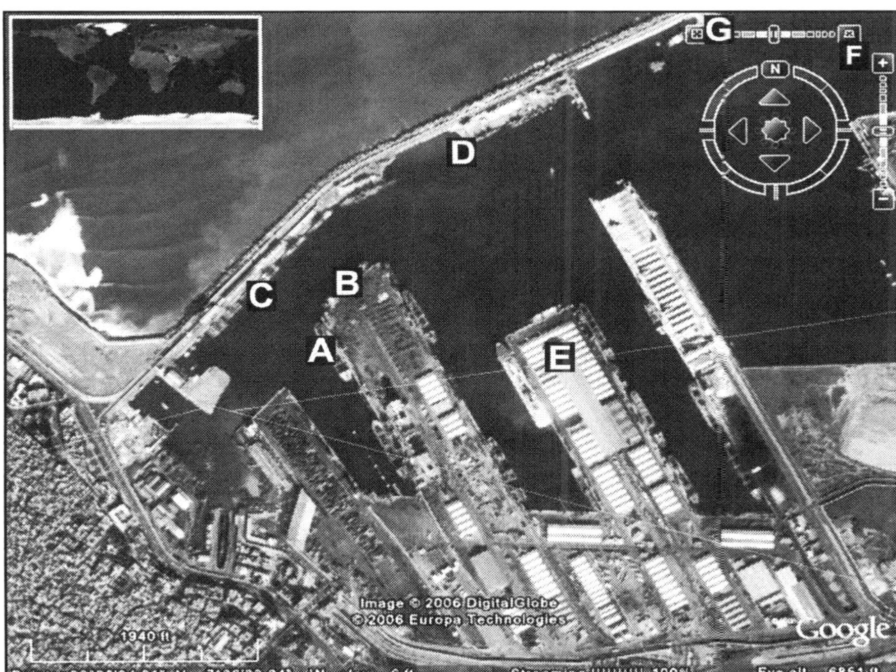

Figure 1.24. A satellite picture of Casablanca Harbor with locations indicated as follows.: The Jean Bart *(A); the* Serpa Pinto *(B); submarines (C); destroyers (D); the pier at (E) had not been built in 1942; outer harbor (F); breakwater (around G). (Satellite Image Source: © 2006 DigitalGlobe, © 2006 Europa Technologies, and © 2006 Google.).*

They sent a tugboat to tow us into the inner, "forbidden" harbor. They moored us at position B, with our stern to the bow of the pride of the French Atlantic Fleet, the very modern battleship *Jean Bart*, moored in position A.

Morning light revealed why the inner harbor was closed. Not only was the *Jean Bart* there, but also eight submarines (in area C), as well as five cruisers and some destroyers (in area D). The remains of the Vichy French West Africa Fleet were assembled in Casablanca, and we could see every detail. (Much later, this made for riveting discussions with the United States Navy intelligence officials in Baltimore.) And unbeknownst to us, the Allied attack on French North Africa was planned for November 8, less than two months later.

Before we left Casablanca, a remarkable event ensued. Several hundred men, women and children embarked on the *Serpa Pinto*, mostly families. We found out that these were refugees from the Spanish Civil War that had raged from 1936 to 1939. Of course, they were Republican fighters and their families. The French had taken them in from Spain in 1939, and sent them to North Africa. They came to Casablanca to emigrate to Mexico, and were to come with us on the *Serpa Pinto*.

Figure 1.25. Jean Bart *after the Allied attack (The* Serpa Pinto *had been moored just to the left and of the bow of the* Jean Bart, *pointing up in picture). Image source: Photographers Mate Third Class Bill Wade from an aircraft of USS* Ranger, Casablanca, French Morocco.

The next day we left, but not without one more scare. As the ship was moving out of the harbor, a French police boat approached very rapidly. The police came aboard, grabbed a man, seized his camera and took the film away from him. They had observed him taking pictures of the harbor, and reacted accordingly. I was glad that I had decided *not* to try to take pictures there!

Onward into the Atlantic

As we pushed into the Atlantic, the weather turned bad and the trip felt long; the Casablanca episode had spooked all of us. A few days after we left Casablanca, we heard a strange radio report: The Germans claimed they had sunk our ship. It was a bit unsettling, but it had the usual German accuracy.

Life aboard ship developed some pleasant aspects. A number of the passengers were musicians, and formed a quartet that played chamber music many days in the lounge, which I found very enjoyable. (Once I fell asleep during the music, and had the most pleasant sleep.)

And one day, during fairly heavy weather, we saw a German submarine, on the surface, hurrying eastward. It was rolling extremely badly because of the rough weather. The crew must have been seriously seasick.

Soon American patrol aircraft, large flying boats, called Patrol Boats Reconnaissance, PBYs, began shadowing us, and suddenly we had arrived at Bermuda, where the British had established their main inspection station in the Atlantic. We were escorted in by a British corvette, a small escort warship, which was rolling so much in the large waves that we could occasionally see partway down into its funnel. This was my first experience of "Bermuda Triangle" weather, but not my last.

When we disembarked, the British took all our papers and books. (They returned all the material a few days later.) While there, I passed many an hour talking with a junior British naval officer, who guarded our ship. He had barely survived the sinking of his ship in a naval action, and was stationed in Bermuda to recuperate. I remember discussing some of the British naval losses, such as the battle cruiser HMS *Hood*, then the world's largest and most beautiful warship. It was suddenly destroyed after a very brief battle with the German battleship, *Bismarck*. He explained that the German gun sighting equipment was extremely accurate, but fragile and vulnerable to shock. Thus, the Germans were often able to make their first few salvos hit their targets, but once their own ships had been hit, their optical gear was out of alignment, rendering their return fire inaccurate.

After a few days, our ship was allowed to sail, and went directly to the Florida coast.

And I had my first sighting of pelicans! (An ugly looking thing,that I became very fond of later, in my many tropical trips.) After we had passed Florida and were out on the open sea, we were suddenly threatened by an American bomber. It circled overhead, lower and more menacing by the moment. Our ship sailed on. Finally, the aircraft made a very low pass with its bomb bay doors open, and we saw two large bombs hanging right over our heads. At that, the ship finally stopped.

We stayed in place for several hours with no idea what was going on. Much later, I learned that we had been headed directly into the American fleet that was being assembled to take the American troops to North Africa, for the invasion in November. Our skipper had ignored a request to stop. As a result, we had almost stumbled into the invasion

Figure 1.26. A sunny day on ship boat deck. CH personal collection.

Figure 1.27. Spotting Florida. CH personal collection.

armada, which would have blown the highly secret preparations for the attack on North Africa. Finally, we resumed our course to Cuba.

Cuba and Mexico

We arrived in Havana, Cuba, to warm Caribbean breezes. By this time we were bored, but had at least begun to feel safer now that we were well inside the protective perimeter of the United States Navy. A number of refugees disembarked, and we set off again the next morning.

A few days later, we docked in Vera Cruz, Mexico, a large and busy harbor. The Spanish Republicans and other passengers disembarked, and a number of Mexican men came aboard to inspect us. Some wore police or army uniforms, but most were civilians – yet all of them carried .45 caliber Colt pistols or other guns. I had never seen armed civilians before, but realized it was another introduction to the New World.

Finally, we departed for the final leg of our long journey, to Baltimore. At long last, we were going "home."

Figure 1.28. Regina and Karl, taken ca. 1945. CH personal collection.

The End of the Beginning

In Baltimore harbor, the process of admission began on board the *Serpa Pinto*. All of us, including Betyar, underwent thorough medical examinations. Immigration came next, and then discussions with Naval Intelligence officers, who were intensely interested in our experience in Casablanca Harbor. They quickly realized that I had seen a great deal, and that I knew quite a bit about ships, so they spent some hours with me, and asked for some more detail about some matters.

They asked how I knew which battleship I had seen. I told them I had noticed two large ship screws sitting on the quay, with *Jean Bart* written on them in large chalk letters, so I assumed that the ship was the *Jean Bart*, and she had only two of her four screws fitted. They inquired as to whether the ship had steam up. I told them I had observed a small plume of white steam coming out of the funnel, suggesting that the ship indeed had steam up. I also mentioned that only one of the two large gun turrets had guns installed. And I explained how some of the submarines in the harbor conducted submerging exercises right where they were moored – and that the subs were loading torpedoes.

Figure 1.29. Home at last!
CH personal collection.

I still had no idea how very timely this detailed information would be.

These discussions went on for some time, and actually kept us on board an extra day. But finally, on the evening of October 11, 1942, we were done. The officials stamped our passports with the next day's date, as we were expected to disembark on October 12. We were able to telephone Uncle Karl to say we would see him and Regina the next day.

And then, the powers that be reversed themselves, deciding there was no need to keep us after all since we had successfully immigrated. They sent us off the ship that very evening instead of making us wait until the next day. They didn't have to ask us twice and we didn't look back. We disembarked, hailed a cab, and headed to Uncle Karl's friend's apartment – only to find it dark and empty.

We sat down on the stairs to wait, tired and hungry, luggage at our feet. For a moment, it seemed our journey was still not at an end. But not long after, Karl and Regina arrived, and were surprised and thrilled to find us on their friend's doorstep. After being told we would not arrive until the following day, they had gone out for dinner. We must have looked as hungry as we felt upon hearing those words – because within a few moments, Karl and Regina shepherded us to a little sandwich shop around the corner where we had our first American meal. We were "home" at last.

Postscript to the Trip

On November 8, 1942 – just a few weeks after we arrived in Baltimore – the Allies landed in North Africa. The fighting in several places was bitter, and the Vichy French Navy fought with determination. The *Jean Bart* was seriously damaged.

Earlier that year, the United States Navy had won the critically important Battle of Midway, a win that changed the course of the war in the Pacific. Soon after our arrival in the US, the British chased Rommel out of Egypt and Libya, and he never returned. The Russians had stopped the Germans at Stalingrad. The tide had finally turned.

As Winston Churchill said at the time: "It is not yet the beginning of the end, but perhaps it is the end of the beginning." And the new movie "Casablanca" was being shown in the US. (It bowled me over, the impact of the great events of the day was powerful, and the atmosphere in France and in Casablanca at that time were shown perfectly.) I felt that for me, in many ways, it was also the end of the beginning.

There was so much more ahead. Next would come college, graduate school, and really growing up.

Figure 1.30. Jean Bart's stern hit.

Figure 1.31. Another Jean Bart's stern hit.

Images 1.30-1.32 source: RW Butler. http://polyticks.com/bbma/jeanbart.htm.

Figure 1.32. Jean Bart's bow hit.

CHAPTER TWO

From the Mountains to the Ocean

Childhood

It is the middle of the night, and I awaken from a hallucination, drenched in sweat and burning with fever. My father shakes me awake as I cry, "I was there! We were all going to be killed!"

"Calm down," he says. "What was it?"

"I was on the ship!"

"Whose ship?" my father asked.

"Rozhestvensky's! We were sinking, the ship was on fire!" I told him.

He asked, "Where was it?"

"Tsushima, of course," I replied, still caught up in my dream.

"What you were doing there?" inquired my father, understanding how real the vision was to me.

"I was Rozhestvensky's adjutant," I told him. And for that moment, it was real.

I was about ten years old, and ill, with a high fever. (It was the first hallucination in my memory.) I had been reading a three-volume book about the Russo-Japanese war, and was fascinated by the Battle of

Tsushima. The books were written by Captain Vladimir Semenoff, a survivor of the battle, and I found them captivating.

As I recall it, Semenoff's story begins as he is on the Trans-Siberian Railroad, traveling from St. Petersburg to Port Arthur on the China Sea, to join the Russian Pacific fleet. While on the train he receives word that war with Japan has broken out and that the ship he was supposed to serve on had been sunk by the Japanese. Semenoff continues to Port Arthur, and is appointed Executive Officer on a cruiser that later manages to escape during a great 1904 battle at Port Arthur against the Japanese fleet. They run out of fuel in French Indochina and then turn to Cam Ran Bay, where both the ship and Semenoff are interned by the French.

Semenoff is released, and made to swear he will not fight against the Japanese again. He goes to back to St. Petersburg, where he is then immediately appointed as an aide to the Russian admiral, Rozhestvensky, to serve on the return fleet. You see, the Russians built a second fleet to go out – a rescue fleet – to relieve the siege at Port Arthur. And this fleet was annihilated by the Japanese in the Battle of Tsushima in 1905.

During that battle, Admiral Rozhestvensky is wounded and taken off his flag ship by a Russian destroyer, along with Semenoff. Later, when that destroyer surrenders to the Japanese, Semenoff tries to kill himself, believing (and with good reason) that the Japanese will execute him for violating his previous pledge not to fight against them. A friend of his intervenes and keeps him from shooting himself – and as it turned out, the Japanese do not kill him. He returns to Russia, writes his fascinating and influential book, and then disappears from history.

Semenoff describes the long voyage of the fleet from St. Petersburg to Tsushima – which took over half a year – in detail. It culminates in the decisive Battle of Tsushima, and reading his account early in my life, that historical event became a deeply personal event, and drove my hallucination.

These books were important when first published, and made a deep and lasting impression on me. Most importantly, the main lesson of the battles of Port Arthur in 1904 and Tsushima in 1905, that the biggest guns were most effective, caused a revolution in naval architecture and strategy, they confirmed that "all big gun" dreadnoughts were a good idea. Plans of several navies to build "all big gun" ships had begun before 1905, but the evidence of the battles was convincing, the building of the British HMS *Dreadnought*, and of the USS *South Carolina* was begun. *Dreadnought*'s keel was laid in 1905, and she was under way in 1906. They transformed all major navies. (Thanks for help with the details of the timing, Linton Wells!) Looking back, I can trace a path directly from the tales of daring deeds at sea that I read as a child to my lifelong fascination with history, exploration, my appetite for adventure, and my abiding love of the ocean. And if I look even more closely, I can find a common denominator: my father. He taught me to look at life as an endless string of opportunities for adventure, and when I was a child, he was there to make sure everything came out all right.

Some of my earliest memories are of talking about these stories with my father. Sometime he would tell me stories from his life and work. Sometimes I would read others' stories and then discuss them with him. We had a fairly extensive library at home and I worked my way through it as I grew up. As a child, I loved reading about explorers, particularly Arctic explorers. One day my father pulled down a book and said, "Here, why don't you read this? I think you'll enjoy it." It was the story by Fridtjof Nansen, a Norwegian explorer who tried to get to the North Pole in a specially designed ship, the *Fram* (which means "forward" in Norwegian), that didn't quite make it. Nansen ended up adrift for a year or two in his ship.

My father was a Captain in the Austrian Army in World War I, and was interested in all things military, so we read our way through various war and naval stories. The Semenoff account of the Russo-Japanese War was one of them, but I also read books about Admiral Nelson and accounts of the Battle of Lissa, between the Austrian and Italian navies in 1866 (which the Austrian navy won). In fact, my father insisted all his life that a relative of ours named Victor Herzfeld had commanded a small Austrian ship in the Battle of Lissa. I haven't been able to find any documentary evidence of the relationship, but there are two pictures of Victor and the other ship commanders celebrating the battle in the Vienna War Museum. Several friends of mine have seen these photos and agree with me that Victor looks quite a lot like my son, Paul. So maybe my father was right. I think my father gravitated toward naval histories because his first choice before World War I was to join not the Austrian Army, but the Navy. But, one of his eyes was bad from a young age. (I remember that he occasionally wore a monocle on his bad eye, which was quite a fashionable thing to do at the time.) When he applied to the Navy, they said, "Well, we really shouldn't take you because of your eye problem. Of course, we will, if you really want to go, we'll let you. But you need to think about the following: As soon as you make a mistake, everyone will say 'What did you expect? He's half blind.' Do you really want to carry that around?" He decided it wasn't worth that risk, and chose instead to join the army, specializing in heavy artillery, wound up commanding a battery of motorized 12 inch mortars.

Many, many stories came out of his experience in the Army, and the horrors of World War I.

At school I learned about the Turkish wars, when the Turks laid siege to Vienna in 1529 and again in 1683. The first attempt failed, but the Turks came very close to success the second time. My father took me to the places around Vienna where some of the main events took place, and history came alive for me.

And I remember him taking me to watch the military parades in Vienna, a very exciting event for me. He gave me a running commentary, and quizzed me on my knowledge of which parading army unit was which.

I also remember the Feast of Corpus Christi every summer. It was a religious procession, but the military participated as the Archbishop carried the Host throughout the city. (The mixing of the Roman Catholic religion and the military was a characteristic Austrian feature.) High dignitaries followed the Host in the procession, in order of rank and status. My father was highly decorated in World War I, and walked alongside those who, like him, had won the "Iron Crown of Lombardy," a high, very old and venerable decoration for valor. He was by far the youngest of that crowd, something he was very proud of; had the monarchy survived, the distinction would have made us low-level aristocracy.

He was devoted to the Church in what we call the "Vienna Catholic" sense – which means you go to church to be baptized, married, and buried, but not necessarily every Sunday in between. On the other hand, my mother was a Lutheran, so to marry my father in a Catholic wedding, she agreed to a Catholic wedding, and also agreed to have me raised and educated in the Catholic tradition. They were married in the Cathedral in Salzburg.

I have many memories of attending the noon Mass together at the Coronation Church when we lived in Budapest, and then walking from the church to the royal palace to watch the changing of the guards who stood watch over the Crown of St. Stephen of Hungary. It was an elaborate ritual, and we enjoyed observing it together, and listening to the military bands.

After my father's death, I attended Mass on my own. I became much more serious about Catholicism once I moved to the US and attended Catholic University. My Uncle Karl attended Mass every day, and I often went with him. I drifted away from the Church later in life, but will always

Figure 2.1. Austrian 12 inch mortar in Word War 1, ca. 1916, like that which my father commanded. The 12 inches refer to the caliber, the internal diameter of the barrel. CH personal collection.

Fig. 2.2. Father in uniform, ca. 1917. CH personal collection.

be grateful for having gotten a sense of religion early in life.

The Army was the most important thing in my father's life, it meant everything to him. I always believed he lost his main reason to live when the Austrian Army disappeared after World War I. He never got over it; he was an unhappy man for the rest of his life. After 1934, he was accepted in the Austrian Army Reserve, but it was not the same to him. But he loved being able to put his dress uniform back on for those parades and official events. It was a little bit of the old life, at least for a few hours.

Despite his heartbreak over losing his identity in the Army, he maintained a relatively balanced perspective, which he passed on to me. When I was a child, examining his uniform and decorations, I pointed out, "It looks to me like you got big medals for smaller feats, and lesser medals for the really big stuff." He replied, "First of all, that's true. And second, don't worry. If you do enough, it all averages out in the end." It was wise and true advice, and it has helped me ever since.

My father and I did more than just read about and talk about the adventures of others. We had some ourselves.

Many who served in combat in World War I had to take huge risks. My father, like many officers, was one of these. He hated war and the people who started wars, but he did what was his duty. He was not foolish, but he did a lot of things that most reasonable people would describe as dangerous. Luckily for me, he instilled in me that same willingness to take risks (together with care) – the rewards have been well worth it, and it was a thread running through my life.

Fig. 2.3. Father in uniform, with mother and a friend, ca. 1936. CH personal collection.

Vacations and Battlefields

One particular memory stands out as a minor and exciting risk. I was about eleven years old, and my mother, my father, and I were vacationing on a beach in Italy. We were at a place called Lignano, near what had been the Italian Front during part of World War I. Purely by chance, part of the beach was being used for a couple of weeks by an Italian anti-aircraft battery for live fire practice. They deployed on the beach, left the cannons there at night with guards, and during the day shot live ammunition at targets that were towed over the water by airplanes.

My father sought out the commander, a captain in the Italian artillery, and they quickly established that for some short but significant part of the war, they had been on opposite sides of the same part of the front, shooting at each other, trying to kill each other.

Figure 2.4. "Helping with auto racing" ca. 1928, note Rainer von Habsburg in racing goggles at left. CH personal collection.

The natural thing to do was discuss this all over a long lunch and a bottle of wine. That became a number of long lunches, and many bottles of wine – after which they were both in very good spirits. As a consequence of their growing friendship, I got to see the battery up close, many times and, most exciting, I got to shoot a machine gun with live ammunition several times, out over the water, aiming as if it were a fire hose. I was careful, of course, and there was someone standing next to me the entire time, ready to wrestle the gun away from me if anything went wrong. It was exciting and fun. I vividly remember the bullets skipping over the water like stones, a curtain of water spraying up each time.

This wasn't the first time my father had allowed me to join in exciting adventures. After his years in the war, he had two major jobs in Vienna. In the first one, he worked as an engineer designing car engines for an Austrian automobile maker called Gräf & Stift. They made luxury cars before World War 1, including the car that Archduke Franz Ferdinand was riding in when he was assassinated in 1914. (The car is now in the Vienna Museum of Military History, and still has some blood on the seats.) As a side job, my father trained the racing car drivers who drove the Gräf & Stift racing cars. They specialized in mountain races.

My father and mother took me along to a number of the races. They were high up in the mountains, it was bitter cold, and sometimes in the middle of snowstorms. My father always drove the race course the night before the race to assess the conditions on the roads – how all the turns were, what was wet, what was dry, whether there were any mudslides, or other information that might help his drivers. He would drive, my mother would take notes, and I would be in the back seat.

The next day, I was allowed to go to the races. I remember seeing a number of famous racers and being allowed to touch a few of the cars. I also recall being aware of the element of danger. They were, after all, racing at high speed on winding mountain roads – not smooth, flat race-tracks – and occasionally there would be a crash. For a young boy, it was hard to beat the noise, excitement, and adrenaline, of the races.

Flying, an Early Indicator

Watching the car races was just a warm-up for the flying. My father made sure I had the opportunity to fly in airplanes early on – starting when I was about five years old. My first memory of flying is of going up in an Austrian World War I two-seater fighter plane, and I sat on my mother's lap. The plane was then about fifteen years old and basically made of paper and wooden sticks, with an open cockpit. I remember being afraid that if I were to stamp hard on the floor, my foot would punch right through the floor and the entire contraption would fall apart. It was exciting and also terrifying.

The pilot, Godwin Brumowski, was famous throughout Austria. He had been the number one air ace in the Austrian Army during World War I, with the highest number of shoot-down victories during the war. He was highly decorated, like my father, and they had become friends. My father arranged for Brumowski to take me up with him several times.

I loved it, and couldn't wait for more. The next time, I went up with him by myself. The Zeppelin was coming to Vienna for the first time, and

Brumowski said, "Let's go see if we can find it." We flew all around the countryside and could not find it – a lesson in the difficulty of air combat control. It was tremendously exciting and still counts as one of the high points in my life.

In 1932, my father left Gräf & Stift and went to work for the Standard Oil Company of New York, which had an office in Vienna. He traveled frequently – to Turkey, Bulgaria, Romania, France, Italy, Poland. He often talked about his job – keeping an eye on the future fuel needs of the Air Forces of the countries he was dealing with, and also keeping track of what the demand for fuel was going to be. That was why it was so important to get him out of Austria quickly once the Germans took over Austria. But before our relocation to Budapest, I was able to fly frequently in the company planes of the Standard Oil. When my father's colleagues would come from the company's headquarters in their company planes, they would often take me for rides in

Figure 2.5. First flight, ca. 1934, on mother's lap; the pilot Brumowski at right; Karl Georgi (a friend of my father) center, in back. CH personal collection.

their planes. Those planes, of course, had closed cockpits, and usually were single-engine aircraft. The pilots would do many maneuvers with me – flying straight up and then stalling, so that the plane started falling and oscillating, it was thrilling.

Once I became an adult, flying became the norm. I often rode in military airplanes and was able to sit as co-pilot with my hands on the controls. That, too, was exhilarating, and took me back to those first takeoffs with Brumowski. When I was just a child, ascending with him for the first time, I would never have imagined that one day I would consider flying the Concorde across the Atlantic at super-sonic speeds just part of an ordinary day.

Finding the Ocean

But my most challenging and satisfying adventures have taken place not over the ocean, but in it. This may surprise most people, they don't think of Austria as synonymous with the sea. But the Austria of my father, the "old" Austria – had a major maritime life, a substantial Navy, marine commerce, and plenty of coast, harbors and beaches. And Austrians basically came in two flavors. The majority were mountain people – the skiers. Then there were a few outliers – the ocean people. I grew up to be one of those. I never took much to skiing – though I did learn to ski. I didn't like it and wasn't very good at it. But in the ocean, I always felt completely at home.

Again, I trace this back to my father. I realized only recently that almost all of our summer vacations were organized so he could go to the beach with us, as well as visit the old battlefields he had been on during World War I. So we spent a great deal of time around the Adriatic.

My earliest memory of the ocean is being on the Riccione beach, near Rimini, Italy. I was about five years old, and the sea struck me as just one giant bathtub. I asked my mother, "Can I go in that big bathtub?" She said yes, and in I went. The water at the beach was shallow, got deeper very gradually, and so I just toddled out farther and farther, until I decided to sit down in the water and play.

I also remember my father teaching me how to build volcanos in the sand. He made a big cone of sand, maybe two or three feet high,

Figure 2.6. Going out to visit Italian battleship Duilio, *ca. 1930. CH personal collection.*

drove a hole to the center of it with the handle of a shovel, and then hollowed out a little cavity at the bottom. He burned some paper in that cavity, and out of the top of our "volcano" came real smoke! It impressed me tremendously, though it turned out much less impressive when I tried it with my own children many years later.

One year we vacationed near on the Adriatic, and the town was visited by an Italian battleship, the *Duilio*. We went aboard it as visitors, and I was really impressed by it all. Not realizing then that I would later in life spend a lot of time visiting warships.

I learned to swim when I was about six or seven years old in Vienna, in a swimming pool, complete with floatation devices, coaches, and lifeguards. Not long after, we vacationed in Lussin Piccolo, on the Adriatic coast. I remember how my mother and I raced to the open water. Without thinking or worrying, I jumped in alongside my mother and started to swim. The water was deep and so clear you could see every stone, every fish. All of a sudden my mother exclaimed, "Hey! Do you realize this is your first time swimming in open water?" She wanted to make sure I was fine, and I really was. I realized I knew how to swim, and just kept going.

And really, I never stopped. Swimming became my only real sport, and it led to my two lifelong passions: sailing and later, scuba-diving.

The swimming led naturally to boats. I have loved boats for as long as I can remember. On our sea side vacations, my parents took me out on local fishing boats, often on overnight trips, and watch bringing in the fish in the morning. A quick breakfast of freshly grilled fish, raw mussels with a squirt of lemon, and some red wine became a new type of feast for me.

Still later, when I was married and living with my family in New Jersey, I had a small catamaran sailboat. We would take it to a nearby lake, and once up to Cape Cod. It was fun, but not what one would consider serious sailing.

Much later, after my divorce, I met a woman who had a sailboat in the lower New York Harbor. She invited me to sail with her, which I did. Eventually we chartered sailboats in the Caribbean, with a professional crew, and took kids and friends along. And then finally, we bought a boat together. It was a 26-foot boat, with cabin and inboard engine, and in that vessel I learned to be a reliable, confident sailor.

Sailing Solo

As it turned out, I realized I enjoyed sailing alone more than any other way. The beauty is that there is no one there to interfere when things are going well. You feel like the king of the world, and you don't need to depend on a soul for anything. But on the flipside, there is absolutely nobody to help you if something goes wrong. And something always goes wrong sooner or later.

When sailing solo, your every step on the boat has to be planned. Every potential problem has to have an alternate solution; and when you move in the boat, you must have several plans where to step, in case your foot cannot land where you wanted it to. You have to have two answers for everything, and you have to have an out for every move you make. You must wear a good flotation device, and you must be safely tethered to the boat. Otherwise, you lose your boat or the boat loses you; you may or may not come back.

About that time I met Zofia Dziewanowska, an elegant and accomplished physician. We became good friends, and after a year we became engaged and then married. We had a happy time, traveled and sailed together a lot. But after several years, serious problems arose which turned out not to be fixable. We were divorced in 1988. It turned out to be a very sad period of my life.

Several times, I chartered a boat in the British Virgin Islands. The sailing and diving there was usually spectacular. But you had to be careful. Once there was really bad weather ahead so we reserved a buoy at our destination (marine radio is great) and arrived early. There we were, watching as the whole anchorage filled up rapidly. After a while all the buoys were used, and many people without buoys anchored in the sand. This is generally fine unless the weather is bad, and this time, it was. In the middle of the night, many boats that were just on anchor were drifting past all the moored boats. A lot of boats crashed into others, and so that night we had to fend off drifting boats; there was not much sleep.

Another time, when I was sailing alone in New York Harbor, all of a sudden I realized I was rather rapidly approaching a telephone pole that was floating almost completely submerged. It was exactly lined up with the boat. I was making 10 knots, a pretty brisk clip. Had I kept going, that telephone pole would have gone right through the boat and sunk me, but I managed to steer around it.

I avoided sailing at night – it's just too hard to see, and too risky. Once at night, I almost ran over a tugboat towing a barge. The barge did not have the mandatory light, and I could not see the barge until I was right on top of it, but I was ready with a panic maneuver, and it turned out all right, but it scared the hell out of me. So I always try to get home while there is still daylight.

I very much understand the desire of those who have sailed around the world. My sailing was on a much smaller scale, but I was interested in every detail – from keeping my engine going to fixing the toilet. When the sailing is good, and you are in control, there is no feeling like it. You

Figure 2.7. Sailing in Wellfleet Bay, Mass., ca. 1969.
Getting boat ready, with Tom and Paul. CH personal collection.

Figure 2.8. Sailing in Wellfleet Bay, ca. 1969, with Paul.
CH personal collection.

Figure 2.9. Sailing in the Grenadines, in chartered boat, with Jim, our skipper, sailing instructor, and chef. CH personal collection.

are engaged in a sport that uses your head and body completely. The physical and mental challenge is nothing short of exhilarating. I miss it; but I know I'm now too old to sail solo safely.

However, I'm not too old to enjoy the sea, to be calmed by it, and to be reminded of the many life-lessons I learned on the water, the most important of which is this: If you set the boat up properly, it can handle itself to a large extent. You can kick back a bit and enjoy the ride. But you must stay alert. You have to look at the water all the time, because there are undoubtedly surprises ahead.

But the great discovery, a true life-changer, was scuba diving. In 1967, when I left the government, for what I thought would be a more leisurely life (silly me), I toyed for a while with the idea of taking up flying, which I had so enjoyed since I was a boy. As I studied its potential for an action sport, I quickly realized, that to be considered a safe pilot, I would need to fly about 200 hours a year. Between work and family, I just didn't have the time to give to it. Since I was already snorkeling with the kids during trips to the Caribbean, I thought: "Why not scuba?"

So in the late 1960s, I took what is called a "resort course" in scuba, which was essentially one day of instruction in the swimming pool, and a brief outing. It was deeply disappointing, the course could not get me properly into it. I wanted to do it right – and this wasn't it. I wanted the entire experience.

Diving

And then came the miraculous year of 1970, about which more later. I became involved in a study for the Defense Science Board, and was sent to San Diego for two weeks to chair the ocean surveillance summer study. Two weeks in San Diego meant three weekends and if I was ever going to learn scuba, that was the time and place to learn.

I called a diving school in San Diego called the Diving Locker, which was then considered the Harvard of American diving schools. I explained the constraints of my schedule and they were happy to work around it; mornings, evenings, and on the weekends. My instructor was a fellow by the name of Flip Nicklin. He was a young kid then, and has since become a world-famous underwater photographer who has authored and contributed photography to absolutely gorgeous books, especially about diving with whales. He was the best. The course was difficult, demanding, and a fabulous experience. When I passed the test, I knew it was just the beginning. And I've been diving ever since, though recently it has become more of physical challenge than I can handle.

It is physically and mentally demanding. And with the high levels of complication, precision, and concentration comes a tremendous payoff – a feeling of serenity, exhilaration, and self-confidence. It allows you to be completely, intensely focused, yet also detached. It's difficult to describe, and is more than a little addictive.

It appeals to me for many of the same reasons sailing alone did. There is, of course, the risk. You have to do it right, or you are in trouble. If you are diving alone, there is no one down there to help you. Every decision counts. You have to know exactly what you're getting into, but also be prepared for surprises.

Learning to dive in San Diego, I was quickly involved in diving through kelp, which was fabulous. Kelp beds are unusual things. Kelp are actually one-celled plants yet typically 80 feet long, with three different tiers. Down near the ocean bottom, kelp resembles the trunks of trees in the forest. You can swim between them with ease. And at that depth, the water is often clear as glass, and you can see fish, abalone, and all types of sea life for a distance of 100 feet. A little higher up you will get into the "branches," but they are few enough that you can still swim through them – and amazingly, you will see completely different kinds of fish.

And then there is the top. With "leaves" lying flat on the water, it appears from below like the canopy of a forest. When you want to get to the surface, you look up to see if there is any free spot in the surface without kelp. You aim straight for that place, and with care you will come up there. But kelp is tremendously strong. You cannot break it, you cannot pull it apart. You can cut it with a knife, but that takes too long. So if you do get caught in the kelp, you absolutely must remember to go backward – never, ever go forward. If you go forward you will wear yourself out. People drown that way.

But while the kelp beds can be treacherous, they are fascinating habitats of sea life, and they have one advantage: the big sharks rarely go into them. So if there are great whites around – and there often are in California – kelp is the safest place to be.

I've had no encounters with sharks during my California dives. But when I went diving in Fiji, we were always surrounded by sharks. They would come near when we came up to the surface near the boat. After a while we noticed that they would never be in front of you, but always behind you. We would signal each other to say "You got one!" And when you would turn around, you'd be face to face with a large, toothy shark, not five feet away. But I have never had trouble from an aggressive shark. I've seen lots of sharks, but they were either uninterested or more scared than I was.

Once I was diving with two of my kids, Tom and Paul, in the Cayman Islands. We were going along a canyon in the coral, when all of a sudden a really large white-tip shark (who are notorious aggressors) came over the shallow reef into our coral valley. We looked at him, he looked at us and then promptly swam as fast as he could into deeper water and was gone. We didn't stay around to see what happened next – we turned around too, and admitted later that we were breathing just a bit harder than normal after that encounter.

Another time, in Fiji, the well-known dive instructor Dee Scarr led a group of us to dive there. It was quite magic. And on a dive in a relatively shallow part of the reef, about 50 feet down, her husband Dave had a video camera set up on the bottom, on a tripod to catch the action. After a while, I saw a shark approach from quite a distance, over shallow coral and down toward us, and I took various pictures of it. But then it decided to come straight at me – exactly at my depth. I snapped a quick photo when he was about 20 feet away but then decided I was not going to get into an argument with him. So I inhaled a little bit and rose about 10 feet higher, and he went below me. Later, Dave was furious with me. He said, "Damn it! You spoiled a wonderful shot! I was trying to capture what would happen when you two met, and you blew it!" I apologized, but it is hard to feel genuinely repentant about avoiding a close encounter with a shark.

In my experience, if you do not bother them, they will leave you alone. Of course, this is not always the case. There are some truly dangerous sharks, like great whites. If you are in a black wetsuit they will take you for a seal and sink their teeth into you. Once they realize their mistake (that you are neither seal nor sea lion), they will probably spit you out, but by that point your arteries are severed and you may not make it. They are seal-eaters, not man-eaters, but it's not like you have a chance to show them your ID card. Speaking about sea lions, I once had a fabulous encounter with a California sea lion off the coast of Catalina Island. The dive boat was anchored on a seamount just a cross from Avalon on Catalina. And the sea lion started playing with me. She would rapidly swim toward me at high speed, then veer off at the last moment. When I thought she was gone, she came up to me from behind and nipped my flipper with her teeth. This went on for a while, but when I saw a male sea lion look at us from not far away, I decided this had gone on long enough and slowly swam back to the boat.

Barracudas are another supposed threat. Once, during a dive in the Caymans, I cut my leg deeply on some coral. I was bleeding profusely and my dive buddy and I headed back to the boat. But between us and the boat was a very large barracuda. He smelled the blood and bent into a zigzag, with his front end swinging back and forth like a door. We recognized that as the attack mode, but fortunately were able to avoid him and get back on the boat fast.

Another time, I was diving in the Bahamas and taking photos. A huge barracuda approached and all I could think to do was stretch my legs out as far as possible. He swam right between them, and I even managed to get a blurry photo from above it.

Figure 2.10. Diving with Wes Andrew, from a Navy boat, ca. 1974. Wes had several encounters with great whites. CH personal collection.

Apart from the risks posed by sea life, there is also the very real risk of mechanical failure, or of something unexpected going wrong.

I once went diving on the Palancar Reef, at the southern end of Cozumel, Mexico. It is a gorgeous reef, but dangerous because it is both large and deep – starting at about 100 feet and going down from there. During our trip, there were strong currents – too strong. We were taken way down, then way up, and much too fast. Ironically, none of this posed a danger at the time. Instead, we ran into trouble when, at the end of the dive, my group surfaced and the dive boat was a mile away. Fortunately, the sea was flat as it can be. We blew our emergency whistles and shouted, but the boat remained where it was for quite a while – the crew had fallen asleep. When they finally came, they were not at all apologetic. Instead, they blamed us for getting lost. While it wasn't the ideal way to end a dive, the scenery was worth it.

Another time I had a close call, I was down at about 80 feet, on top of a wreck off the New Jersey coast, and my breathing equipment failed. My son, Paul, was at the other end of the wreck and did not realize there was anything wrong until I was on the surface. So I made what is called a swimming ascent. You keep breathing the little air you have in your equipment, over and over. But it was scary; I went up faster than I would have liked, and was thinking the whole time of the danger that too much nitrogen had dissolved in my blood, creating bubbles that would be the end of me. It's for exactly that reason that diving alone is not recommended. (But still, I like to do it.)

I love the silence; when you dive, you can usually only hear yourself breathe. It's a very rhythmic, calming sound. Sometimes you hear ships on the surface – big, low, rumbling sounds. Sometimes you hear parrotfish chomping the coral, and if you are very lucky, you might hear a whale. I once had the great luck of listening to a humpback whale singing off the Australian coast. Research on humpbacks has shown that whales have complicated songs, and different pods of whales have different dialects, and different groups make unique and highly structured sounds. The human ear cannot distinguish it of course, however, electronics can capture the details, and you can sense that it is a primitive language.

The underwater world is one of magic, and you can experience it, see it close up, but not touching it. Perhaps the most wonderful and mysterious experience is to dive with a "bait ball," a school of fish perhaps thirty to fifty feet in diameter, the fish, small, like sardines, are tightly packed into a ball of fish swimming very rapidly, all in the same direction, making the ball seem to rotate. There are perhaps a hundred thousand fish in the ball, all swimming defensively, because they are being herded by large fish on the outside of the ball, like tuna or jacks, or sometimes dolphins or sharks. You can swim through the ball since the fish avoid contact, and in the process you make a visible hole in the ball. I have taken many pictures of this and many minutes of video. It is an extraordinary spectacle of feeding and survival. And you can observe it closely, swim with it, and through it.

Diving with giant Manta rays has also produced very memorable experiences. They are huge, some with wingspans of over twenty feet. They are related to sharks, among other things they share the lack of a skeleton, but unlike their cousins, they are never aggressive. However, they are curious and come to inspect divers all the time. They may come very close if they want to. With their wings flexing and wing tips flapping, they look much like flying birds – majestic and awesome.

Diving also allows the mind to drift, while still concentrating intensely – much like sailing. I've had some of my best ideas underwater. You never race underwater. You push ahead gently; it is more like a controlled glide. And you can let your mind wander a bit to ponder all kinds of things – from concrete questions about family, management problems, and scientific questions to more existential questions about the meaning of life. It's hard not to ponder larger questions when you are surrounded and absorbed by the most arrestingly beautiful life. Everything you see, everything you touch, is alive. Even dead coral provides a home for tiny, gorgeous fish. I live in the ocean, not just on it. I'm most myself when I'm in the water. I still think of particularly memorable dive trips when I want to fall asleep, immersing myself in the atmosphere, remembering the feel of the water, the sound of my own breathing, and the sight of all that natural wonder. Diving deep in Bonaire produced a wonderful surprise, Garden Eels. Small worm-like eels burrowed in the sand, by the thousands. They are shy but very curious, like most marine creatures. When you come close, they hide, but after a while they come out to look. They are wonderfully peaceful, and thinking about them in bed, has provided many times of peacefully falling asleep.

That's what I like best – the silence, the solitude, and the self-sufficiency. For a half-hour, maybe an hour, you don't hear, see, or depend on anyone except your dive buddy, who for the last twenty-plus years has been my darling wife, Shannon, who has become an excellent diver. She has been my boon-companion as well as dive buddy, sharing the glorious experiences of all of life!

I feel so fortunate to have been able to share this passion with her, with my children and stepchildren. Like me, they've become good at it, and in the process, they have discovered things about themselves that most people never learn. I think one of the problems of the modern world is that many people have had it too easy; they never know what it is like to be on their own. We do our children a disservice if we shield them from risk, instead of letting them discover for themselves, "Will I panic if I'm by myself and a shark comes straight at me?"

Shark attacks are an extreme example, but I was brought up to be independent, to think strategically, to be prepared to handle the unexpected – and the often unwelcome – on my own. Otherwise, I might have been paralyzed by fear that day in Vercelli when I realized my mother and Betyar were not on the train. But instead, I had the

Figure 2.11. Shannon and my stepdaughter, Rose Shuman right, and Rachel Shuman, Left in Curacao airport in transit to Bon Aire, ca. 1991. Photo by CH.

presence of mind to gather our possessions, get off the train at the next station, and wait on the platform for my mother to catch up. It was simultaneously an act of faith and of reason – the same instinct that led me slowly to the surface of the water, decades later, when I realized my breathing equipment had failed.

This attitude was an objective of my upbringing. My parents, like all parents, were not perfect. But, they made a conscious choice to expose me to situations where I would, quite literally, sink or swim. And I am grateful for it. I think back to my father's willingness to send me up alone in an open-cockpit plane before I was even ready for school.

They knew, as I have come to know, that in real life, there are big, difficult things that you have to do on your own. Sometimes they come upon you suddenly and there is no time to seek advice. Sometimes they are so deep and difficult that you cannot even seek counsel. You have to learn to solve your own problems, to rig your own sails, to make your own luck. Because sometimes, we really are on our own.

Figure 2.12. Garden Eels, Bonaire, at dive site "Something Special." Photo by CH.

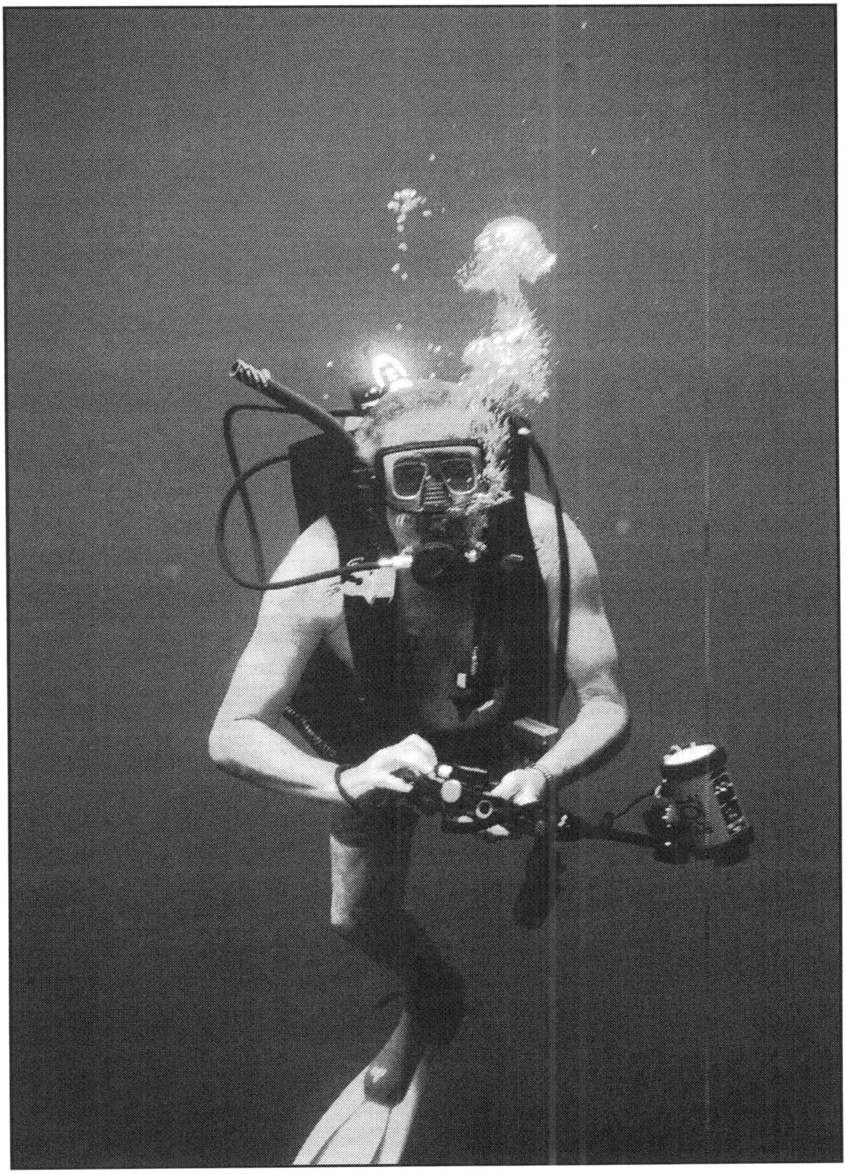

Figure 2.13. Author with camera, photo by Shannon Herzfeld.

CHAPTER THREE

A VERY LONG, BUT VERY BROAD EDUCATION

First, A European Education

Coming of age in wartime Europe resulted in a rather patchy formal education – one that spanned schools, cities, even the sea. Before the war forced my family out of Vienna, I went first to a Catholic school, the so-called Piaristen. It was run by a Catholic teaching order, and the school was very prestigious. I did well there for four years, and had wonderful teachers. And I learned to enjoy learning. It was a great gift for life. Then I attended the junior high school at the Piaristen for a year, but I did so poorly in Latin they would not keep me. This was considered a disaster in my family, because that meant I could not go on to the Gymnasium, which required Latin and Greek, and was required to be accepted into the Vienna university (a family disaster because most of the men in the family had gone to the University, for several generations).

After that, I went for a couple of years to the public school, a so-called real-gymnasium. (It had Latin, but not Greek.) Many prospective engineers, lawyers, and businessmen attended. I felt I learned relatively little there. And then we left for Budapest, I went to what was also technically a real-gymnasium. As a result I received a good education in Latin, which turned out to be extremely helpful, as Latin is the key to half the European languages. It was tremendously useful later in life. And the basics were taught very well; not until when we were ready to leave in 1941, did the education become Nazi-oriented. And I learned to navigate public transport on my own in Budapest, good practice for my budding ability to speak Hungarian. This was all secondary to the biggest excitement, though: the classes were co-ed! This was my first regular contact with girls, and was quite exciting, and produced many surprises. The most dramatic was that many girls were really smart.

My fathers' death in January 1940 changed everything. We had to plan our escape, described in Chapter 1. My education was on this track until it stopped abruptly in June 1941; I was sixteen then, so I was in the middle of high school by American standards. I knew much more about some things than high school students in America, like history, geography, chemistry, a lot of mathematics, and languages – German, Hungarian, Latin, French, and English. But my social skills where minimal. But all that changed when we left Budapest. I tried to supplement my education by reading everything I could get my hands on, from history to science fiction.

Portugal and Baltimore

Once we made it to Portugal in the summer of 1942, I went for two months to an English "public school" – which means that it was exclusive, expensive, and private – called St. Julian's. It was run for the local British community and their families. At that time about half the students were, like me, refugees in various stages of leaving Europe. It was a very demanding education, I was officially studying for the matriculation exams at Oxford and Cambridge, the subject matter was incredibly hard in some ways. For example, the school required a minute understanding of English history – I remember that with horror. But I began to play sports, like boxing, track and cricket. And I learned to mingle with kids my age again, this time in English. But I was lonely, a bit of an oddity and somewhat of an outcast.

Then we arrived in the United States, where I didn't quite fit anywhere in the educational system. In early 1943, I went for two months to a school called Baltimore Polytechnic (I was living with my mother who had a job in a bookstore in Baltimore); it was one of the two great high schools in Baltimore at that time. I went there mostly to get me used to American schools, which were very different from the schools I was used to. Here, I again was subjected to an unpleasant (and to me, irrelevant) history class: a detailed American history course about the time just before the Civil War. I remember hearing about Senator Clay, and not knowing who that was or how any of it fit, or why I should care about it. There was no one there to tell me. (I learned about him and his importance much later in life.) Most of the kids had absolutely nothing to talk to me about, and vice-versa. I had trouble with the language, I had trouble with the culture, I had trouble with the war, which was very real to me, but for most in America at that time was still a very distant matter. But I dipped into the high school to get a feel for the American system, and despite the culture shock, I'm glad I did. And I learned about machine shop, and one of my triumphs was to make a chisel, from a metal rod to the finished hardened steel edge. After a couple of months, I entered the Catholic University of America in Washington, DC, and moved from Baltimore to Washington.

Catholic University

After high school, Catholic University was the best option. It had very few students because by that time, mid 1942, most the high school kids had been drafted, so the University took everyone who had any credentials. Now, I had no credentials, but they interviewed me, and agreed to take me on probation. It helped that my uncle was chairman of the Physics Department. I had to pay tuition, and was able to rent a room nearby. Finding a room was a lucky break because Washington was crammed full of people who had come to the city to work on the war effort. But it turned out that my uncle's half-brother, Herbert, was just then drafted into the Army. He was a graduate student at Catholic University in the Music Department, and so I inherited his room. The timing was quite miraculous, and the room was an easy walk to the university, and also walking distance to where my uncle and aunt lived at that time. (After the war, Herbert became a successful orchestra conductor, doing most of his work in Europe, he died quite young.)

My university experience was a mixed bag; it was better than I had expected, and not as satisfying as I had hoped. The first problem was that I wanted to become an aeronautical engineer, at least initially; I wanted to build airplanes. This isn't hard to understand, considering that the war was getting to be a big air war. But the Aeronautical Engineering Department at the university was just a bare-bones thing at the time, because everybody had left for the war effort. And so I started out with a program of Aeronautical Engineering and took a lot of engineering courses, like surveying, engineering drawing, and shop. However, the freshman chemistry course I took changed all my plans. It was taught by a brilliant young man named Alfred Sklar, and his course

Figure 3.1. My class in Engineering, ca. 1944: Strength of materials, Prof. Frank Biebermann, center, author in front, second from left, George D Edwards, a good friend, next to Biebermann. Photo by Harry Rauth, of Hagerstown, MD. (He was listed in the 1940 census, and was then 72 years old. I remember him vaguely.) CH personal collection.

turned me toward a lifelong affair with the physical sciences – physics and chemistry. So that became what I concluded I really wanted to do.

The university administrators helped me get there. They said, "The way you can save all your engineering credits is to switch your major to chemical engineering, and then you can do any chemistry or physics courses you want. There are a certain number of chemical engineering courses you have to take, but all your engineering credits, that otherwise would have been blown, will count toward graduation." And so I did, taking three full semesters every year, which was quite a stretch.

The Chemical Engineering Department was then a part of the Chemistry Department, which was small but excellent. The head of the department was Frank Rice, a gifted scientist and a wonderful teacher. I throve in the environment. It turned out that I had to take one engineering course in chemical processes engineering, at Johns Hopkins in Baltimore, which meant commuting by train one night every week, for one long semester. It was a hard course, but fascinating and fun. I had a great professor named Bonilla, and I learned a lot, much of it useful in many later challenges. I remember that in those days, the trains were full of soldiers coming or going; there was often no place to sit except the floor of the train car.

An extra-curricular activity came my way: The student news paper, called *The Cardinal*, which had the usual college news, sport, parties and such. But Paul Twiddy, then the editor, asked me to write a weekly column about the war, commentary, not reporting. It was a sort of op-ed page. I enjoyed doing it and it was not too hard. Few people talked to me about it, and it was rated several times the "least read" feature in the paper.

The Army wanted to draft me in 1944, but I failed my physical, I had serious tuberculosis. There was no real treatment then, all one could hope was that it would get better. At that time it was considered a sentence to a short life. I was worried whether I could expect a full life. Writing this at the age of eighty-six, I know that I was extremely lucky. Finally, I graduated *cum laude* at the end of my final summer term, in 1945.

Next was the challenge of graduate school, what to study, where to go, where to apply? First, I applied to Princeton. They wrote me back a very nice letter saying, "Well, we would love to have you, but we can't, because by statute, we have to have a certain ratio of graduate students to undergraduate students. We are not getting any undergraduate students because of the war, so we cannot take any new graduate students." This was a perfectly plausible reason, no one knew when the war in the Pacific would end. Then the Catholic University chemistry department advised me apply to Harvard next. So I did, and I was accepted with a full scholarship.

I was set to begin in the fall of 1945, but it was not to be. In August, the United States dropped two atomic bombs on Japan, ending the war. Right after that, the newspapers reported that Fermi and Urey and Teller – and many of the other greats from the Manhattan Project – were headed to the University of Chicago. Over dinner with my uncle and I said, "Karl, I want to go to Chicago. All the great people I want to study with are there." He said, "Well, try it. What can go wrong?" So I wrote to Harvard turning down their acceptance – which was a huge leap in the dark – and wrote to Chicago, asking for admission. They replied, "Well, we have no idea who you are but your grades look all right. Why don't you come on probation?" So I did, and after a quarter at Chicago, I was off probation and officially became a graduate student in the Chemistry Department.

The University of Chicago

Of course, I had no money, but got an lab assistantship that covered living and tuition. I rented a room on the South Side, north of the campus, and lived in that area for the next five years.

I was in the Chemistry Department, but was in a very "physical" branch of the department. For the first year I had a very straight chemistry curriculum. The second year I got into chemical physics, and began the transition to physics.

My first mentor at Chicago was James Franck, who won the Nobel Prize in Physics in 1925 for work he did between 1912 and 1914, which indicated that Bohr's model of the atom (electrons orbiting around the nucleus, in discrete energy levels at very specific energies) was fundamentally right. James Franck was in the Chemistry Department but he was a physicist. During the war, Franck ran the Chicago portion of the Manhattan Project, where the first nuclear pile (reactor) was built. At the end of the war he chaired a committee that produced the "Franck Report," which famously recommended that the bomb be first used in a demonstration, and not on a city. The advice was ignored, and Hiroshima was the first target. When I got to Chicago, Franck had become interested in the scientific problem of photosynthesis: How do plants turn sunlight into sugar, and into cellulose? He explained to me very early on that there is this problem of taking a big quantum of visible light – having a lot of energy in it – and turning it into a lot of small steps, because all the chemical steps involved are really small in terms of energy. And nature somehow discovered that mechanism; he wanted to understand what the mechanism was. It was kind of a bizarre subject at that time, but he was one of the people who solved it.

Besides being a famous physicist, Franck was also a good friend of my uncle's. I had met him in Washington in October 1942, when he came to visit Karl. So, on my first day at Chicago I went to see Franck in his office. As we were talking, in walked Harold Urey. Franck and Urey were old buddies. Urey says, "Hi Franck, I'm here. I arrived today." Urey was also in the Chemistry Department. The spirit at Chicago was that you did not worry too much about what the stuff you did was called; what mattered was that you did interesting work.

Figure 3.2. James Franck, ca. 1950. University of Chicago.

In those days the Chemistry Department consisted of three big pieces. The largest, and most famous at that time, was organic chemistry led by Professor Kharasch, a brilliant organic chemist. The second was inorganic chemistry – an old subject that had gotten a huge boost during the Manhattan Project, which discovered all kinds of new inorganic chemical systems related to uranium, plutonium and their fission products, as well as boron, which was an essential part of the fission system. The third – and eventually the most significant branch – was physical chemistry/chemical physics. And that is where I wound up.

I started courses in chemical physics in my second year under Joe Mayer, who was at Johns Hopkins before the war. I did very well, making the second highest grade in a very large class in that very demanding course.

That year, Enrico Fermi taught the Nuclear Physics course at Chicago for the first time. The class was very large, with many people who became famous later. They had Nuclear Physics before, but nobody like Fermi, and I was allowed to audit that famous course. One of the great things about the university was it was very easy to audit most anything in most any department, provided there was space in the lecture room and that you did not make a pest of yourself. I also audited Clyde Hutchison's and Edward Teller's Chemical Physics courses, and had a marvelous time. I had taken it from Joe Mayer before, whose course was very rigorous, and carefully structured. By contrast, Clyde's course was rigorous but narrow, and very clear, but Teller's kind of roamed all over the landscape, but was wonderful, a true inspiration. When he lectured, he rattled off complicated concepts extremely quickly, while simultaneously writing on the board. We often wondered what the connection was between what he wrote down and what he thought, and we decided that he could talk ten times as fast as he could write, and was thinking ten times as fast as he could talk. So the connection between what was in his head and what was on the blackboard was sometimes remote. If you did not know what he was talking about, you did not understand very much. On the other hand, if you knew the subject and his style of thinking, which was always very paradoxical, with lots of intellectual surprises, you could have a wonderful experience, and learn a lot. He was a marvelous teacher, and later became both a colleague and a friend.

Fermi had a totally different style. He was quiet, thoughtful, and systematic, with a prodigious memory. He used to say that he never had to do an experiment over that he had already done, nor repeat a calculation of anything he had already done. After he died, the notebooks he had kept during his professional life were published. They were meticulous, recorded with precise dates and details. And the last volume was the index for the whole, which he had kept up his whole life. It showed he was not bragging, he had everything he had done written down in his notes, clear, concise, complete. A prodigious feat that impressed all.

Significantly, Fermi is the only man I ever knew who was a great experimenter, as well as a great theoretician, a great teacher, and a very nice person. Also, he always went after the big questions, the big ideas or big problems; indeed, he delighted in finding the solutions, or encouraging others to do so, and in this way he contributed so much to science, in a magisterial style. One of the best examples I remember is his influence in figuring out how to date organic objects using carbon-14 radioactivity. Willard Libby and others had discovered that carbon-14 is weakly radioactive and has a half-life of about 6,000 years.

One day in a seminar, Fermi said to Libby: "Bill, you really ought to see if one can make a clock out of that. You ought to then to be able to use this to determine the age of things that were made from wood or cloth or something like that – things with carbon in them." And, by God, Bill Libby actually did it. His name is now synonymous with radiocarbon dating; and he won the Nobel Prize for Chemistry for this work in 1960. His work transformed archaeology and history, yet it sprang from what Fermi considered to be a semi-casual conversation. He was not going to do it, but he wanted somebody to do it – and Bill Libby was the resident expert in all the science that was required.

That kind of conversation was commonplace at the University of Chicago, which made for a heady, exciting atmosphere. And the ability to audit courses or drop in on lectures or seminars made it that much more intellectually stimulating, particularly because the professors were always happy to talk to students.

Harold Urey is another example (I audited his thermodynamics course). He was the one who first said, "You've got to go to the moon, got to look at samples there, to see how the earth was made." I remember once going into Urey's office, to ask him about something, and saw a huge photograph of the moon – maybe 5 feet diameter – set up there in his office. He said, "Herzfeld, (in those days, people called students by their last name, before the modern age where everybody is on a first-name basis with everybody) help me understand something. Let's look at the moon here. Some of these craters look oval and some look circular. The ones near the equator look circular; the ones north or south of the equator look oval. Are these circular or not? Let's calculate it." So we sat down and calculated a few dimensions in simple geometry, and yes, they were circles. He said, "Now I know for sure. Thank you."

I once asked him: "Professor Urey, how is it you know so many fields, if you do not mind my asking?" He said, "It is simple: I try to learn a new field every month." That advice stayed with me. Later, when I was at ARPA, I tried to do the same. I asked my staff to give me reading matter, some books and scientific papers on particular topics; on one trip the target was geology, to read on the long airplane trips, and came back understanding geology pretty well. And I needed to understand this because of nuclear test detection issues ARPA was working on, and I really wanted to understand the scientific basis of the arguments in favor of reliable detection of nuclear explosions. We were about to change geology with our large aperture seismic array. I wanted to understand the consequences of that, and I got it, thanks to Harold Urey's advice.

Another major influence was John von Neumann, the famous physicist and mathematician, who once came to Chicago in the late 1940s to give three lectures on electronic computers, which had just started to work on real problems. I had studied von Neumann's great and important book on Quantum Mechanics, and thought it was fabulous. So when they announced these three seminars, I asked my professor and thesis advisor, Clyde Hutchison, whether I should go. He said, "No, don't bother. It's nothing related to what we are doing." I went anyway – and it changed my life.

Von Neumann viewed computing as a radically different new key discipline. The old meaning of computing involved dozens or hundreds of people using mechanical calculators, and later electric ones, to laboriously do it all by hand. (There was some electronic calculation done at Bletchley Park and Fort Meade during World War II.) But the Electronic Numerical Integrator and Computer, or ENIAC, was probably the first programmable electronic computer. It was built during World War II for the Aberdeen Proving Ground. During his lecture, von Neumann talked a little bit about the ENIAC, whose use he had supervised, and I was fascinated by his lectures.

Von Neumann predicted that programmable electronic computers were going to be very powerful. One could do straightforward calculations that would enable one to solve problems that are too complicated to do any other way.

For my thesis work, I was in the business of finding the solutions to matrices that arise from the forces on the atoms I had measured in the lab. I wanted to do more complicated problems, so solving matrices was critical. It is hard to approach doing complicated matrices by hand. The computers did it accurately and very fast. Von Neumann opened a new door, another way to do science. Science used to have two branches – doing experiments, or doing theory. Most people did one or the other, a few did both. But now all of a sudden there were three, the new thing: scientific computing, which is not quite one or the other. He had such a unique view of it all; it transfixed and changed my outlook, and me.

Yet, the ENIAC was integral to what became my biggest failure at Chicago. Joe Mayer asked me to assist him on a problem of statistical mechanics, using the ENIAC to calculate the following problem: What if you had a big box full of billiard balls, and they all get set in motion? How do they move? The idea was to follow their motion, collision by collision, and see if these balls behave the way we think they should. I worked on that for three months and got nowhere.

There were several problems. I did not know enough classical physics to do it and I found that programming in machine language for computers was totally hopeless for me. There were no computer languages yet, so you had to write out all the instructions in "machine language" and that was quite beyond me then. It was horrible, so I did not do it. And I did not know enough to ask for help. Well, Joe Mayer was rather angry at me for that, and it cost me. But I deserved it. A few years later Bernie Alder solved this and other problems in molecular dynamics computation. He received the Presidential Medal of Science a few years ago.

But I walked away from Chicago with several important things.

One, I got an excellent top-down view of science – all physical science and some mathematics – that gave me a good nose to look for the fundamentals in any big problem. What makes a problem tick? What does a good, serious solution have to have to make it work? How does the universe work? How do pieces of it work? Never mind the details, never mind about how you catch the fish, but what kind of fish is it? How does it swim? How could you catch it?

This has been useful for me ever since. To this day, I skim scientific journals and can tell quickly whether it is worth for me to read it or not, because I can swiftly get the real scientific point being made. Usually, I do not bother with the details of how; I do not need to know that. And if I do need to know it, I know how to learn it. The big changes and significant facts stand out like lighthouses. And those are worth reading.

Another important thing I learned was from a seminar Clyde Hutchison ran for three or four years with a bunch of us, all students. Once a week, we would get together after dinner and work for four hours until everybody was tired. We took the great, modern physics texts, one at a time and "beat them to death."

We checked everything. We did every problem, and double-checked every derivation, and discussed it all as a group. We started with JH Van Vleck's *The Theory of Electric and Magnetic Susceptibilities,* and moved on to von Neumann's book, *Mathematical Foundations of Quantum Mechanics,* then Condon and Shortley's *The Theory of Atomic Spectra,* and Tolman's *The Principles of Statistical Mechanics*. I still have these basic books, and refer to them occasionally.

What I realized is that you can learn anything if you work at it. There are true big mysteries in the sciences, like the connection, if any, between quantum physics and relativity. While many people have attacked this problem, and there hints of solutions, the deep issue is still quite mysterious. And there are lots of *small* mysteries in the sciences, they tend to be matters of detail, and are perhaps things you do not want to spend time on. You can think anything other people have thought. You may not discover it, but you can follow in their footsteps. And you do it best in a small group with very bright people who really want to understand the subject.

And the seminars were a great way to get to know people. One of the remarkable people I met at Chicago this way is David Mann, and his wife Vera, and we have been connected professionally and personally ever since. We will meet them again.

But perhaps the most important thing I learned was something about me: I was making it at the University of Chicago, so I was pretty good at this stuff. But there were a few people in my cohort who were vastly smarter and better organized than I was; they all had Nobel Prizes within 10 years. So I learned at an early age that I was good in science, but not absolutely top-notch. And this is a bitter and sobering, but useful, thing to understand about oneself, early on. But I found later that I had greater talent in organizing scientific attacks on large problems.

But though I may not have been the best, I made it through. My thesis was about the magnetic susceptibility of five different compounds of uranium. They were Uranium (4+) with four electrons removed, which left two electrons in an empty shell. The controversy at that time was whether these were d-electrons or f-electrons; two units of angular momentum or three. Several students of Clyde Hutchison and I proved unambiguously that the actinides – uranium, neptunium, plutonium – had f-electrons, and that these elements were a second row of rare earths in the periodic table.

The thesis and result were easy to explain. I determined the magnetic susceptibility of five compounds of U+4 by measuring the force on samples in a magnetic field, at different temperatures. The strength of the force at different temperatures gives the magnetic susceptibility, which tells you how many units of angular momentum are carried by each of the electrons. That, in turn, tells you where on periodic table the element belongs. I measured the force on several samples of five different compounds of Uranium, each with a different crystal structure, at several different temperatures, at room temperature, at dry ice temperature, at about minus 30° Celsius using liquid ammonia, at liquid nitrogen, and liquid and solid hydrogen temperatures. Liquid hydrogen boils at 20° Kelvin and solid hydrogen at 13° Kelvin. I performed the experiment for each of five different uranium compounds, with different crystal structures, to make sure that the results were consistent. And they were.

As a side note, I took my youngest son, Paul to the University of Chicago during his college search, and introduced him to Clyde Hutchison. Clyde was a great host, thrilling for Paul. Clyde also was good showman: Reaching for the bookcase behind him, Hutchison asked Paul, "So, would you like to see your Dad's note book for his thesis? Here it is." Paul looked at it with awe. I shuddered a bit, glad to be done with it.

While I was working on my thesis, I started to teach physics at DePaul University in the Chicago North Side. They were trying to build up their science departments, and offered me a teaching position in physics, even though I was still a graduate student at Chicago. It was a lot of work, but I learned a lot, I needed the money, and it turned out to be a lot of fun. The students were few in number but high in quality and enthusiasm. The department wanted me to start out teaching a basic course in atomic physics. There were a lot of textbooks, but I did not like any of them. So I decided to teach from the literature, using the scientific papers from Maxwell and Faraday in the 1850s on to Bohr's papers in 1911, and Einstein's first papers on special relativity and the photoelectric effect. Over the year we read perhaps ten or fifteen papers, but very carefully, those by Maxwell and Faraday on electricity and magnetism, also Crookes' papers on inventing the Cathode ray tube (by the way, he also found X rays but did not understand it, he merely warned readers to protect fresh photographic plates from being near his tube when it was turned on, and wrap them in lead foil, because he explained they fogged, where exposed, when they were near his tube when it was turned on). I assigned several papers at a time, and then we would discuss each in turn in class. It was quite successful in that the students got a whiff of what doing physics was like in the late 1800s and early 1900s, and I learned an awful lot that was not usually covered in standard courses.

My social life in Chicago was mostly based on the Catholic student club at the University. There I learned a lot about philosophy and religion, mostly from fellow graduate students in those departments, whom I met at the club. I became a regular writer for a Catholic student publication, called *Concord*, dealing with social and philosophical issues, based at Notre Dame University in Indiana. That too, widened my horizons.

Figure 3.3. Teaching at DePaul University, ca. 1950. (Note the board had not yet been cleaned from an earlier French class.) CH personal collection.

I finished my thesis in March 1951, and handed it in to be professionally typed. The next day, I was drafted into the Army, and headed off to basic training. The Army checked my health, and surprise, surprise, the tuberculosis had healed by itself, and my lungs were fine. So, next came going into the Army to add to my education. I went, with a train full of new soldiers to Fort Riley, Kansas, where I became an infantry man.

My final exam at Chicago had been set for June. Hutchison, not understanding about the Army, said to me, "Maybe we can waive the exam so you do not have to come." And people who had served in the military told him, "Clyde, don't be silly. He'll be dying to get a long weekend in Chicago. Make him come and take his exam." So I did come back from my basic training in the Army, and took the exam.

James Franck was the chairman of my committee, and there were a number of other professors, including Joe Mayer, Bill Libby, and, of course, Hutchison. There was also a professor by the name of Herbert Brown, yet another future Nobel laureate. He taught among others, a legendary course called Organic Analytic Chemistry. (We used to call him "Red Brown in a Blue Suit" because he had red hair and always wore blue suits.) He insisted that absolutely no one could get a PhD from the department without taking this course in Organic Analytic Chemistry.

Figure 3.4. Getting a PhD in uniform, at the University of Chicago Graduation Ceremony, August 1951. CH personal collection.

Well, I decided early on I would try and not take his course, and then encountered him on my final committee. I knew I was in trouble going in, but I knew what questions he usually asked, one in particular: "How do you tell primary amines from secondary amines from tertiary amines?" It was a very important practical question in analysis of organic compounds, and I studied it until I understood it exactly. So sure enough, when it was his turn to question me, he said, "Mr. Herzfeld, how does one tell apart primary, secondary, and tertiary amines?" I gave him a really detailed, unhurried, thorough answer. He said, "Thank you. No more questions." And that, mercifully, was that.

The United States Army

Going into the Army brought good news, that my TB had healed, but otherwise there were lots of down sides to it. I was determined to make the best of it. I boarded the train to Fort Riley, Kansas, with many other young men. The Korean War was in full swing. The North Koreans had overrun South Korea in June of 1950, and pushed the small and unprepared American force back into the Pusan Perimeter. Then, in the fall of 1950, General McArthur led the Battle of Inchon with a brilliant flanking attack in the north of South Korea. His forces trapped and destroyed the North Korean army. He kept going north, and was near the Chinese border when a large Chinese force chased the US out of North Korea.

By March 1951, we were mostly fighting against the Chinese. The joke at the basic training center was, well, we will be here for a few weeks and they will airship us to Korea, because they are desperately short of trained people.

That was the backdrop for all of us entering the Army at that time. It was sobering to have to prepare oneself for the likelihood of being sent off to battle on the other side of the world. And I had quite a bit of culture shock to boot. This was the first time in my life I had the opportunity to get to know really average American kids. I was most taken aback by their language; I was shocked by the sheer number of "F-words" per sentence, and the creativity used in combining them.

They may have been crude, but there were sterling qualities about those kids. They were all younger than I. Half of them were volunteers for the paratroopers and were 17 or 18 years-old; some were younger but had lied to get in. At just under 26 years-old (the limit of the draft age) I was the oldest, and the only one with a PhD. Some of them may not have been educated, but they had guts, and a tremendous willingness to do the right thing.

I realized immediately that I had to be accepted by them. I had to persuade them that I was real. It was a challenge, but I made it. It helped that I turned out to be a good shot. And it also helped that I knew from my father how the military system worked – contrary to what many people thought, the army was not stupid, just different. The way to get ahead in it was to follow it, while the youngsters wanted to fight it and sometimes got trampled on pretty badly.

My PhD initially made a big splash. Lots of people wanted to meet me, but my company commander cooled that off in a hurry. In an all-hands meeting he said to us, "What we do here is serious stuff. There is a real war on. At the end of this tour, most of you are going to Korea, so pay attention. Your life is at stake." And then he said, "I hear a rumor that there is a PhD among you. Let me tell you that does not count for anything here. My PhD is this," he continued, and pointed to the ribbon of the Silver Star, a high decoration of the Army for valor in combat, which he had been awarded in World War II. "So that is why I run the place and I'm going to run it my way. So is everybody clear, anybody confused?" Nobody was. That is what leadership is about. He bit the bullet fast and all the way through, and that was that. The topic was off the table, much to my relief.

Basic training was a wild experience. We started out by learning to march, and to assemble and disassemble M1 rifles. The M1 was our standard infantry weapon from World War II and was one of the best in the world at that time. It was semi-automatic, so firing each bullet required a trigger pull. When people complained about that, the answer was: "Well, it can be fired as quickly as you can aim it." And the Army's doctrine at that time was "aim to fire." You do not spray bullets, you aim true.

That turned against the Army because the recruits who went to Korea believed that and acted accordingly. But the Chinese had automatic rifles, which were Russian-designed World War II rifles – an excellent design. The kids in Korea came back from combat situations not having fired their rifles at all. When asked why, they said, "I could not see any targets."

So to counter that, they changed the training. They gave us a brilliant demonstration during the summer, when the corn was man high. They put about a hundred cardboard targets on sticks in some cornfields, so you could not see the targets when you were walking through the corn. They lined my company up in a straight line, gave us ammunition and said, "Just put your rifle on your hip, just walk through the corn field, and keep shooting straight ahead."

We did. When we came out, they collected all the ammunition and walked us through the corn. Every one of those cardboard targets had been hit – many times – demonstrating that planned mass fire is very, very effective. They said, "Do not just shoot when you see a target, shoot whenever they tell you to shoot." And that was a good and powerful lesson.

All our instructors had just come back from Korea, so I saw up close what combat does to soldiers. I understood my father better, who had been in combat for four years. They trained us well, and we learned as much from their stories as from their instruction. I remember one guy in particular, a machine gunner, who told how he had fired continuously for several hours during a Chinese attack. The Chinese got so close they tried to grab the barrel of the machine gun. He and his crew survived, but it was tough fighting. He persuaded us that it was incredibly hard – but that it could be done. So I became a very careful recruit because I did not know where I was going next, and wanted to improve my chances of coming back alive.

It was very hard, physically, but after a while I was in the best physical shape I have ever been in my life. We walked a lot, twelve miles a day was average, we carried a lot, weapons, ammunition, tents, etc. It was hot and very demanding. And because the Chinese were skilled at fighting at night, the Army added night training for us, at first taking it out of our sleep time. We were so tired that during lecture breaks we all went out and fell asleep on the streets. Once, I was next in line for the machine gun firing, and fell sound asleep for a half an hour right next to the firing machine gun. It was fired the whole time, but I slept like a baby.

My goal was to make it with these guys, and I did. I won fame by inventing a game with the tough guys. We would stand in the open meadow and throw bricks at each other, underhand, and catch them without gloves. Some people cut themselves and some did not; my hands grew very strong and I did well with the game. I was able to walk longer than most of the guys. I was able to carry the heavy part of the machine gun and I was quicker in the arithmetic you have to do (mostly involved with estimating distances and such).

Also, I was a better shot than most of them. We fired all the standard infantry weapons. I remember one day, we were shooting prone on the firing range with the M1 machine pistol, a leftover from World War II. They put cardboard targets on sticks and I blasted the cardboard off the stick, and kept firing at the stick and cut the stick down a few inches at a time. A large amounts of ammunition was expended, but I did it. Unbeknownst to me, while I was having fun, the company commander came up behind me and said, "Herzfeld, they need you in Korea, get your ass over there." I said, "Yes, sir."

We also learned to use 81mm infantry mortars, and also recoilless rifles, which I had not known anything about. The recoilless rifles were an invention of World War II, which are designed in such a way that a soldier can fire an artillery-sized shell from the shoulder without getting knocked down by the recoil. They are the loudest things I have ever known. Once you fired a shot you were totally unable to hear or even feel where you are for an instant, but they are wonderfully accurate. I took two or three shots at a tank that was a mile away and hit it every time. All of it was instructive; it showed me I could do this stuff.

Being skilled with these weapons boosted my confidence in several ways. It helped me gain acceptance with the young kids, who were very gung-ho and very big risk takers. And it also helped me interact with the noncoms, some of whom were not nice people.

There was one major incident with a corporal whose name I have forgotten, who was seriously hounding and harassing me all the time – for weeks. One day we were on the range with live ammunition, learning how to shoot out of a bunker in a defensive mode, against notional attackers. He was giving me an exceptionally hard time. After a while of this, I had finally had it, so I stepped down from the stand I was firing from, with my loaded rifle, pointed it directly at him and said, "You do that once more and I will kill you." I meant it, too; I was past caring.

Silence fell. An officer in the bunker called "Cease fire, cease fire!" The whole company ceased fire and they took all the ammunition from our company, marched us back to the barracks and confined us there for the rest of the day and night. The next morning, the corporal was gone and I was still there.

Another confidence-builder came with one of the last tests of the company, called the infiltration course. It was a very elaborate scenario: The troops start out in a trench that was deep enough that your head was below the top. When you climbed out of the trench into the open, you had to crawl under and through barbed wire, and you had to keep your body below three feet off the ground because you were advancing against machine guns that were continuously firing live ammunition over your head. They were arranged in such a way that they could not shoot you, unless there was a major malfunction – but once in a blue moon, someone would panic, stand up and get killed. You had to crawl to, and then past the machine guns that were firing over you, then you had to load your rifle and shoot at targets ahead of you. And to add another degree of difficulty, the exercise was done at night, and it rained heavily that night.

When they gave the order to get out of the trench, everybody stood up and for a second did not move. But after a moment everyone went over the top. It was raining hard, and the mud was like oil. I was about halfway through the crawling part, when the machine gun post right ahead of me started the following game: every time I moved an arm or leg, he would fire a burst over my head. So I knew he was watching me and he was out for me. I was furious, but knew it wasn't personal. Still, I understood how if you are in combat, you really would want to kill the guy shooting at you.

I got through the infiltration course, and I'm glad they trained us with live ammunition. To this day, it is still easy for me to tell the difference between blanks or real bullets from their sound.

Midway through basic training, the Kaw (Kansas) River flooded. We knew there was a flooding danger, so we had already packed up our gear and stowed it on the second floor of the barracks – and a good thing we had. In the middle of the night, we awoke to shouts: "The dam is in trouble!" Several companies, including mine, tried to repair and strengthen it, using sandbags at first. When we ran out of those, we filled pillowcases with wet earth. And the wet earth was so heavy it just tore the pillowcases. We bunched them up on the dam and this worked for a while, but then we saw the water seeping through the middle of the dam. Then they quickly moved us back, and then evacuated us. The barracks were ruined; we never slept there again.

Instead, the whole regiment moved to field tents in the field on the hillsides. Life got a lot simpler; the experience added a degree of realism that was really quite interesting. But it also meant basic training lasted longer than usual – almost a double course.

And at the end of the training, most of the men were indeed sent to Korea. But when my orders came in, I saw that I had been assigned instead to Aberdeen Proving Ground, to the Ballistic Research Laboratory. I was relieved, of course, but I felt a sense of satisfaction in knowing that, had I been sent to Korea, I would have been a good soldier – as good as the German soldiers had been. I had spent years immersed in a flood of propaganda about the German military successes, which was very intimidating. Their last gasp was the Battle of the Bulge at the end of 1944, which caused tremendous concern that the Germans would snatch victory from the jaws of defeat. For seven years, during my most impressionable age, I watched this monster unfold. It was supported by the ability and guts and strength of the individual German infantryman – something that was both frightening and impressive. But through my basic training, I found out what they could do, I could do. That was good to know; it still is good to know.

I was happy to be headed to Aberdeen. I spent my two weeks of leave in Chicago, packed up my belongings, and headed toward Baltimore.

Aberdeen Proving Ground, Ballistic Research Laboratory

I will never forget reporting to Aberdeen. I arrived at the barracks in the middle of the day with my duffel bag and gear, and was greeted by one Willy Widule, the soldier who was assigned to take care of the barracks that day. Willy had a master's degree in mathematics. I introduced myself, and he said, "Great. Do you like classical music?" I said, "Yes, sure." He replied, "Well, I have a surprise. We have a secret." The secret was an extra clothes locker – one that we were not supposed to have – and it was filled with a hi-fi set and records.

So Willy said, "Why don't we put on some music? Do you like Mozart?" He put on Mozart's Piano Concerto No. 21, which at that moment became and is still my absolute favorite piece of music. It was heaven. Army life up to then had been hard and dirty, brutish and uncivilized. And suddenly, there floated down this heavenly music. It was unforgettably wonderful, and marked the turning of the next page in my life's journey.

The Ballistic Research Laboratory was then the best Army laboratory, scientifically speaking. It was strongly influenced by the Johns Hopkins Physics and Chemistry Departments, and was the place for developing everything from the smallest infantry weapons to the largest artillery the US Army had, and everything in between. The testing of all the guns also happened there, as well as the testing of trucks, tanks, and other equipment and weaponry.

I was put into what was called the Interior Ballistics Laboratory. Ballistics divides naturally into Interior Ballistics (what goes on inside a gun), Exterior Ballistics (what happens to the bullet as it flies) and Terminal Ballistics when the bullet hits something. Interior ballistics was the most scientific and complicated area, and the most interesting to me.

The task of modern interior ballistics is to understand completely all the processes that happen inside the gun when you fire it. When the propellant powder burns it pushes the shell out of the barrel, and the shell goes wherever it will. The idea is to predict how far it would go, preferably from first principles. (The great Galileo began his work in physics trying to solve this problem. He failed, but then began the discovery of basic mechanics, so ballistics has an honorable, now largely forgotten, pedigree.) It wasn't hard to write the equations for all of that, but it was too complicated to solve analytically because it involves a highly nonlinear system of equations.

Joe Frazer, who was Director of the Interior Ballistics Lab, suggested I tackle a thorny problem – one that had, in fact, been suggested by my former professor Joe Mayer during World War II. (During the war Mayer advised Aberdeen a lot.) The problem was to estimate the resistance to the motion of a bullet in the bore of a gun quantitatively, and to make predictions about it for different designs and materials. What causes the resistance? How do you calculate it? Mayer had suggested a simple approach: Suppose the bullet is rammed into the gun barrel and there was nothing pushing it. It would stick; everything is solid. Now, what happens when the force of the propellant when it is burned, pushes the bullet?

Mayer's suggested solution was that the force of friction between the bullet and the inner wall of the gun heats and melts some material of the rotating band, that goes around the outside of the shell, and the work done by the burning powder keeps a thin layer of the rotating band material molten. (Rotating bands are usually made of bronze or copper.) That liquid layer lubricates the motion of the bullet and provides a thin film of liquid as it goes down the barrel. Can one calculate how much energy it takes to keep the film molten as the bullet goes through the barrel? And the beautiful thing – *the* lucky break in my professional life to that date – was I was able to solve it and exactly, from first principles. It turned out that the solution was a complicated integral equation, and I had no clear idea what to do with it. But one of my mathematician colleagues at the Lab, Peter Treuenfels, knew the exact solutions to this type of equation. So I could use these solutions to examine in detail what the theory meant, it agreed with experimental data for all kinds of rotating band materials.

It turns out that this is one of the classical problems in the theory of interior ballistics, and I had solved it. It was published as a classified report (now unclassified), and for a while I was famous in that arena. But more importantly, Joe Mayer was very pleased, and forgave me for screwing up at Chicago. I had redeemed myself in his mind, and was no longer a lazy bum.

My days in the Interior Ballistics Lab were really another large dimension of my education. I had learned a field that was new to me. I had received warfare training with all these weapons; now I knew not just how to shoot them, but about the detailed physics, chemistry and mathematics of what went on inside them. I learned about the chemistry of the powder – as it turns out, the shapes of the powder grains are an essential part of the design. There is a whole theory of the shape of grains of propellant to get any pressure profile you want. The barrel has spiral grooves inside that make the bullet spin as it goes down the barrel, and the rotating bands fixed on the bullet make it turn as it goes. You do not want the bullet to tumble when it comes out of the barrel, and having it spin makes it less likely to tumble. Different kinds of materials can be used for the rotating bands – brass, copper, soft iron, steel, Teflon – and all these things were tried, and my theory agreed well enough for each material.

My colleagues were an impressive bunch: several PhDs, several quite gifted mathematicians. They were smart people who had some real intellectual scope. They explored all kinds of ideas, tried all kinds of new designs, built models, and tested them on the range. It was a very rich subject, totally new to me, and fascinating.

Personal life at the lab was also much more enjoyable than in basic training. We had lots of leave, and could arrange three-day weekends. Many others in the company barracks liked classical music as well, and invited me to come hear them sing in the Handel Choir down in Baltimore. Eventually, I auditioned for the choir and was accepted, and making music became very important to me. I sang there for about two years, and it was an incredibly rewarding experience.

When my tour of duty was over in March 1953, the folks at Aberdeen offered me a great job. But by that time, it was becoming clear that classical guns were a dying field, and that the future of everything military was rockets. The Army rocket research was happening in Huntsville, Alabama, and I was not about to go there. Instead, I headed south to Washington, DC, and the Naval Research Laboratory.

The US Naval Research Laboratory

In 1953 I took a job as a researcher at the Naval Research Laboratory, or NRL, in Washington DC. It was the best government science laboratory together with the Bureau of Standards at that time, and the job was

Figure 3.5. My first apartment, ground floor, right, in Washington DC, 1953. Photo by CH.

attractive: I would be able to continue working on problems similar to those I had done for my dissertation. The lab was equipped for that, and they were happy to have me.

But it turned out quite differently than I had expected. The experimental side of my work did not flourish; the lab was more focused on getting immediate results rather than long term experimentation. In addition, I lost interest in the field I had worked in for my thesis, and even lost interest in doing experiments at all.

I do have one memorable moment from the experiment, however. For a while I was continuing the work of my thesis, using a wonderful very large magnet that was powered by submarine batteries in the basement of the building. They were extremely powerful sources of electricity, and I was warned, "Be sure that the only metal objects lying around in your lab are non-magnetic, made of bronze, when you turn the magnet on." Well, when I first turned the magnet on, there happened to be a steel screwdriver on the other side of the lab, and it came flying across the room to the magnet, missing my head by inches. It was an exciting event – but not a reason to continue.

It was not wasted time, though, because I did some good work with a colleague in the Magnetism Section, which I published later in an important paper on group algebras, together with Paul Meijer. Still, it soon became clear to me that I should move on.

National Bureau of Standards

I began to search for a new job, and landed a position with the Bureau of Standards, which was an excellent place, an old and venerable organization that had strength and a charm about it that I appreciated. I had attended some seminars there, and had met the people who ran the Heat Division. The chief of the Heat Division, Ferdinand Brickwedde, hired me, and gave me a title that I have loved ever since. I was the "Consultant to the Chief of the Heat Division, for Theoretical Physics and Physical Chemistry," one of the more grandiose titles I have had. Brickwedde was a wonderful man who narrowly missed the Nobel Prize. His story tells volumes about how science really works.

In 1932, I think it was, he and Harold Urey, who was then professor of physical chemistry at Columbia, decided to try and see why the atomic weight of hydrogen was not quite what you would expect if it was just a proton and an electron. But the weight was a little more. Sure enough, they discovered that a small percentage of hydrogen is heavy hydrogen (then a new atomic species) and its nucleus is made up of a proton like normal hydrogen and also a neutron. It is now called deuterium.

That was a major discovery. Urey got the Nobel Prize, and Brickwedde, who did the experiments, did not – because he was a government employee. In those days, the Nobel Committee did not think that government employees of any country ought to get Nobel Prizes.

Understandably, Brickwedde was an unhappy man for the rest of his life. He was also a wonderful man and a great boss (and eventually I replaced him as Chief of the Heat Division).

I loved my work at the Bureau. One of my most exciting challenges was to work out the theory that explained some of the experiments being done by Herb Broida and John Pelham.

Broida was gifted experimenter in many fields related to spectroscopy of hot gases, and Pelham was an ingenious low temperature physicist, who used liquid helium in large quantities for his research. They proposed the following idea: "What if we generate a flow of gas, send it through an electric discharge, which makes a lot of excited molecular and atomic species, and collect all this stuff on a plate at liquid helium temperature, what will we get?"

Well, they did this, and they got a huge surprise, because the stuff they collected when nitrogen gas was sent through the discharge glowed blue, with bright flashes and little wave fronts of flame going through the accumulated solid. It was all at 4° Kelvin – extremely low temperatures. The spectrum of the emissions from the material was slightly, but measurably, different from that of regular nitrogen, molecules or atoms. I was able to explain the observed spectrum as arising from excited nitrogen atoms, whose normal energy levels were disturbed by the surrounding nitrogen molecules in the solid. (In one of life's odd connections, these experiments turned out to be an early ARPA project. The Pentagon had started to look for solids containing much higher energy than those previously available, for rocket propulsion, and ARPA funded the work. Because my theory worked so well, I had a good feeling for what was going on in cold-stored high-energy systems, and I very clearly understood that there was no possible rocket propulsion fuel to be found in this system. The flashing material did not last; you could not store it except with liquid helium, and even then only for a very short time. It was not exactly what you would use to make a space-traveling rocket. Many people were disappointed, but the facts were undeniable.)

That was some really nice science, and it was widely recognized. While I was thinking about other problems to attack, my career at Bureau of Standards veered toward top management at the Bureau. In those days, the Bureau had twelve to fifteen divisions, the actual number changed as time went on. Division chiefs were big shots at

NBS and in the Washington science establishment; they had very prestigious positions.

I was recruited for management soon after I planned and chaired a symposium on the Measurement and Control of Temperature in Science and Industry. NBS had arranged two of these before, and I recommended a third (because of the many advances made since the last one). It was held in 1960, after about two years of preparation. I urged NBS to cover the older areas of temperature measurement in physics and chemistry and industry, but also to emphasize measurement and control areas which had never been treated systematically before, those related to biology, and that was going to be a huge, complicated subject: For example, how much energy does a cell produce when it lives? How does the production of energy of that same cell change when you raise or lower its temperature? If one could measure this and explain it, this was going to be one of the more important scientific questions of the end of the twentieth century. But first we needed data, and before that we needed to understand how to make the measurements to get reliable data. So there was created a new category in our measurement series; biological systems.

The symposium was a big success. It was well-attended and well-received. I was chairman of the program and editorial committees, and the papers presented at the meeting turned into three big volumes that are still easy to find in any bibliography. Amazon finds them easily. In the process, the top management at the Bureau noticed me, particularly the director, Allen Astin, and the deputy director for physics, Robert Huntoon. Huntoon became my active mentor, and pushed my career along a lot.

Brickwedde was offered a job as a professor of chemistry at the University of Pennsylvania. He had a great career there for many years. When he left the Bureau, Huntoon put me into his place as an "Acting Assistant Division Chief." One of my early lessons from him was that when I asked him too many questions regarding what I should do, he answered, "What are you? You are the Acting Division Chief. So act, for God's sake!" And so I started doing things, and found out that I was good at it.

I officially became Division Chief around 1959 – they "skipped a generation" by appointing me. The grandchildren of my colleagues were the ages of my children. They did it again by promoting Lew Branscomb to become Chief of the Atomic Physics Division. Lew was about my age, and later became Director of NBS. It was incredibly daring at the time, and was a wonderful success, for NBS and for us.

The job was also rewarding personally. I became responsible for the development of the scientific activity of a large group of experienced scientists, and for their professional development. I became responsible for a major standard of measurement (in case of the Heat Division, the national standards for all temperature measurements,

Figure 3.6. Award ceremony at Heat Division, for famous parity experiment, 1957. Many people in this picture were important to me at that time. I wish to recognize Dr. Allan Astin, Director of NBS, at my right; Dr. Robert Huntoon, Deputy Director for Physics, and my important mentor, his head just touches the left border of the blackboard; Dr. Ernest Ambler, in front of Huntoon, then Director of the Low Temperature Section, whose group was being honored, and he became later Director of NBS; in the corner of the room Jim Swindells, Director of the Temperature Standards Section; and Eileen Hanson, just behind Astin, at the time my Executive Assistant. Photo courtesy NBS.

at all temperatures, and all circumstances). When I became chief of a division, several other pleasant things happened: One, I became an adjunct professor at the University of Maryland, with the right to supervise PhD theses there; two, I was a physicist, and I became a fellow of the American Physical Society; and three, I became a member of the Cosmos Club. I particularly liked teaching physics. It was a great deal of work but I enjoyed the interactions with students tremendously. When I left NBS to go to ARPA, I stopped teaching, there was simply no time for it; but I missed the students and their bright young minds.

The Bureau brought me even more than that: a direction for the rest of my career. I learned how to run things. I came to see that doing the technical stuff was relatively easy; the hard part was the understanding of the problem being worked, and seeing a technical and an organizational way to the solution. I discovered that I enjoyed doing this, and was getting pretty good at that. For most any problem that confronted me, I could see how to attack it, and maybe crack it. Often, when somebody tried to explain to me why a particular thing could not be done, while he explained why it was impossible, I thought of three ways to start to attack the problem.

So I took another road, and that has made all the difference (thank you, Robert Frost.). I found that I was better at leading and running things than I was at doing things. I was no Mozart, but perhaps I was a fair Toscanini.

Coming to this realization was bittersweet. It was a bit of a failure, because I had been brought up to think of doing basic research as the ultimate human occupation. But I decided I was not good enough at it to satisfy my aspirations. Whenever one hits that point, one should change to something else because struggling for second best is really hard. Most people are happy if they think they are the best, or near the best, or among the best.

At about this time, on October 24, 1957, I was invited to a meeting at the National Academy of Sciences for a very special occasion: The famous Nils Bohr was the first to receive the newly established "Atoms for Peace Prize" from President Eisenhower. I went and met the famous Bohr, whose key papers I had taught at DePaul University,

Figure 3.7. Bohr receives first Atoms for Peace Award, from President Eisenhower, with Jim Kilian, Presidents Science Adviser. I was in the audience. Source: AIP Emilio Segre Visual Archives.

at a reception, and witnessed the ceremony. Eisenhower gave a major speech about the long range significance of science and technology for the welfare of mankind. A picture commemorates the occasion.

Also participating in the ceremony was Kilian, the new science advisor of the President. At the time I had no idea I would be so heavily involved in the aftermath. ARPA was established soon after, and a few years later I would begin working on the tasks Eisenhower had assigned ARPA. It was another hint from the future.

It was clear that I was doing quite well – understanding the big problems, seeing solutions, finding good people, getting lots of resources, and bringing it all together and make an omelet instead of a mess. And I met David and Vera Mann again, David having taken over a molecular spectroscopy section at NBS, and we continued our enjoyable professional and social connection.

Meanwhile, I became intrigued by a policy problem: It was clear that we, the United States, were going to build very large rockets. By very large, I mean rockets with a thrust exerting a force of the order of a million pounds or more. But to develop these, you have to be able to measure large forces. How do you do that? And who pays for that?

For a while, our space program had one failure after another. Eventually the National Academy of Sciences recommended that the Defense Department put together a joint Bureau of Standards and Pentagon committee, and go around to the major rocket builders, like Douglas Corporation, Lockheed, and others, to see what was wrong with their measurement process. Something clearly was, because they could not do repeatable experiments or tests.

So the committee was put together, I became a member, and we visited many rocket labs and plants to check the chain of taking data to analyzing and understanding the results. We discovered many issues in their science and organization which caused the problem. Reforms were started, and the aerospace industry became a paragon of excellence in scientific and technical measurement.

The focus of the Heat Division in this area was the measurement of high temperatures, because one of the things needed was a rocket engine that put out a lot of thrust. You have to measure the temperature of the gas you create, and the temperature of the rocket motor and the rocket parts to see how hot they get if you are going to design anything that is going to last through what you want to accomplish. This is all part of the grand strategy of designing a rocket.

Well, it turned out there were no temperature standards for those high temperatures. It was my division's job to provide those, and after many months, we did. We learned how to measure the spectra of the atoms and molecules of the very hot gases in a rocket and to estimate their temperature accurately, reliably, and repeatably. A section chief in the Heat division, Henry Kostkowski, perfected this method. I remember that during the first launch of an American astronaut into sub-orbit, we were all in my office listening to the radio report of the re-entry to see if the man inside the capsule came out alive. While we had done unmanned experiments, and we were pretty sure the capsule was safe, but there is nothing like the real thing to get your attention. He was fine, of course, and the rest is history. We had proven we had a practical high temperature standard.

Measuring the force of the rocket motor was also difficult. The way you measured force was to take a cylinder of steel called a force measurement cell and either loaded it with the load that you want to measure (in which case, it gets shorter and fatter), or stretch it by hanging the load on it (in which case, the cell gets longer and thinner). You measure this changing shape by winding a wire around it closely and measuring the resistance of that wire. As the cell gets wider or shorter, the wire stretches or contracts, and the resistance goes up or down.

But nobody had built or calibrated force measurement cells for even a quarter of a million pounds (four of these would measure a million-pound force). There was no program to do that. So I recommended that there should be a national policy that whenever an experiment like the million-pound rocket was proposed, that same program would have to budget and work with the Bureau of Standards to build the measurement standards before doing any big experiments that were likely to fail.

It was a sensible idea, and we wrote it up, to send it up the chain to the Department of Commerce, which ran the NBS. It made its way through the Commerce Department and onto President Kennedy's desk, reaching it – unfortunately, on the day of the Cuban Bay of Pigs Invasion. Quite understandably, he was too busy to read it, and the proposal died. That was life. But as my father used to say, "If you do enough, you'll get enough credit," and so it came to be.

First Contact with ARPA

In May of 1961, as I was sitting in my office in the Heat Division, my secretary, in the manner of the time, told me that I had a phone call from a Dr. Ruina from the Pentagon, and would I take it? It was Dr. Jack Ruina, the Director of the Advanced Research Projects Agency, known as ARPA. ARPA was the big new kid on the block for funding

Figure 3.8. Dr Ruina and I, at 50 year anniversary dinner of DARPA. Photo by DARPA.

large technology projects. I knew it from my early project at NBS, trying to find high-energy solids as propellants for rockets. I was curious to hear what he had to say, and I was also always looking for research money, which was perpetually in short supply.

Ruina asked me to come visit him at ARPA in the Pentagon, which I did. I was treated to two days of briefings on major programs, ranging from Project Defender, the Anti-Ballistic Missile Project, to the effort to help the South Vietnamese combat the insurgency raging there. At the end of these long sessions Ruina asked me to come to ARPA and run Project Defender, a huge program involving almost half of ARPA's budget.

I was intrigued, but after thinking about it for two days, I said no. I was fully engaged at NBS, in a job that was important and that I loved. Well, that was not to be the end of it.

In August 1961, I took a long trip to visit several European Standards Labs. I took the occasion to physically deliver several Platinum Resistance Thermometers to some of the European Standards Labs. They had been carefully calibrated at NBS and were to be used in a round robin, having the same thermometers calibrated at all the labs to get a good consensus about their stability and exact calibration as thermometers at very low temperatures. It was a long trip, partly business, partly vacation for me and my first wife, Norma.

Near the end of our trip we went through Paris to bring thermometers to the International Lab for Standards of Measurement at Sèvres, near Paris. We stopped by to make a courtesy call on the US ambassador, General Jim Gavin, who was a personal friend of ours. At that time France was in the middle of the war of Algerian Independence; the situation was very tense, with lots of police visible, and with occasional violence in the streets. At the appointed time we were sitting in Gavin's office. He was late. His secretary informed us he was with General de Gaulle, the French President, and that Gavin really wanted to see us, and would we please wait. After three hours, he finally arrived, highly agitated, slightly sweaty and disheveled. Cookies, coffee and a little brandy appeared. He apologized, saying: "I have been with de Gaulle all this time; there is a huge crisis brewing. The East Germans are putting up a wall in Berlin. We don't know yet if the Soviets are behind it. They have said that it is to keep spies out; we think it is to keep East Germans in."

The answers to questions about the wall were obvious in a few days. This was a very serious escalation of the Cold War by the Soviets. On our way home I did a lot of thinking. The Cold War had been going on since the mid-1940s, though that phrase was not yet current. I concluded that this wall meant war – not a shooting war, but real war all the same.

When I arrived back in Washington, I called Ruina and asked if the job at ARPA was still available. It was.

I started two weeks later, much to the chagrin of my colleagues at NBS, who felt I had deserted them. But I knew it was the right thing to do, and it proved to be the most demanding and rewarding job of my life.

And so, at long last, my formal education ended. Real life was about to begin.

CHAPTER FOUR

Growing Wings: My Life in and with ARPA

Prologue

My experience at the Advanced Research Projects Agency (ARPA), later called DARPA, the Defense Advanced Research Projects Agency, played a unique role in my professional and personal life – and to some extent still does. It was the most challenging and exciting job of my life. It enabled me to do my best work, with the most important consequences. It was the first (perhaps the only) job that required, and deserved, all my energy, all my ability, all my determination – and made it worthwhile to engage totally. Over time it became a shining star over my life. It also cast some deep shadows. With power and responsibility and action came criticism and opposition; some helpful and well-meaning, some not. Friendships and, alas, enmities, grew that have lasted.

Looking back, I am pleased and satisfied with the results we achieved, and with the recognition and credit we received, although in some instances these were quite slow in coming. The meaning of success and failure became clearer and stands out in sharper relief. Those times and events are only now being seen by others in the perspective we brought to them originally. Old differences are becoming mellowed, at least on my side.

ARPA/DARPA is a singular organization. I know of no other quite like it (though over the years I have searched for similar organizations). And so the ARPA/DARPA story is complicated, mostly because the challenges and opportunities were so diverse and so vast. It is hard to imagine today what was asked of us then, and what was possible to do. The story runs in several streams, so I will tell it as clearly as I can, and hope it comes together for my readers. Life is not like a map, it is more like a carpet, with intricate designs. (Again, this work is not a history of ARPA, but my view of these events.)

Over the years, several histories of ARPA/DARPA have been written. I have found some of these good and interesting, others quite inadequate. Most of them get basic facts right, but miss the deeper context, particularly our intent and the enormous scope and freedom with which we operated. We were expected to solve the problems we were assigned, not just work on them. We were expected to rebuild the house, not just rearrange the furniture, but our masters were realistic about the times required, though were always pressing for results "now".

This report of my life, as is all of it, is based in part on memory, always a somewhat suspect voice, but it is buttressed with a large volume of data from my records and notes, matters in the public domain, many discussions with colleagues, and other sources.

My Introduction to the People and Projects of ARPA

ARPA had started under President Eisenhower, in reaction to the surprise of Sputnik in 1957. It was located in the Pentagon – and in those days you could just walk into the Pentagon and go to any office, if you could find it. I knew a little more about ARPA than most. As I described earlier, I'd had some dealings with ARPA in the late 1950s while at the National Bureau of Standards (NBS,) interpreting Broida and Pelham's experiments with Trapped Free Radicals, a project ARPA had funded. After I joined ARPA, I found out much more. It was a small organization; less than a hundred professional staff. Most were young, highly educated, and technically sophisticated; all were full of energy and ideas. They evidently were able to spend large amounts of money, and had lots of independence, with a minimum of red tape, it seemed they could actually do things, and the tempo of activity was extraordinary. The intellectual demands were a "constant stretch," and the physical demands called for "running at full speed."

The chief projects and objectives originally assigned to ARPA when Eisenhower was President,were as follows: First, fix the space program; this was set up after the Sputnik surprise, it was mostly a task to initiate large technical and organizational moves, to prioritize and to manage; second, do the leading-edge work for Ballistic Missile Defense (BMD) (going beyond the technology of the Army Nike Zeus ABM project that was underway in development, and which had been set up with the intention to deploy a full system); third, to help the South Vietnamese in their war against local Communists; fourth, set up a system for detection of nuclear test anywhere in the world (before this new move, only small uncoordinated efforts were underway); fifth, establish a large university program to reinvent materials science on a modern and interdisciplinary basis, supported by large, long-term funding; and sixth, to make major improvements in rocket technology. Soon after these, the Information Processing Technology Program was also added. Importantly, these were assigned by President Eisenhower, with the advice of Jim Killian, the former president of MIT, and then President Eisenhower's Special Assistant for Science and Technology.

ARPA's Director reported to the number three person in the Pentagon, Harold Brown, the Director of Defense Research and Engineering. The ARPA Director's Office was in the "high rent" district of the Pentagon, on the third floor, in the E (outermost) ring of the Pentagon, close to the office of the secretary of defense.

Literally, the first decision I had to make at ARPA was where I wanted my office to be located. I had about an hour to decide among the three available options: One, very near the director's office (most prestigious), another near most of my staff (most accessible to my crew, but far from my management) or one in between, roughly equidistant from the other two. I saw all three and I chose the middle one, and a great choice it turned out to be – many people were watching and reading great significance into my decision. Most approved the choice. It was trivial in retrospect, yes, but indicative of bureaucratic maneuvering for prestige and access and being observed and analyzed constantly.

The week after I arrived, when I had just begun to get the lay of the land, the second big challenge arrived. Jack Ruina walked into my office with a piece of paper and said, "Charley, you have been given an additional job. The secretary of defense has assigned ARPA a new project, and since you now lead Defender, you are the logical person to run it. We need to start it fast – it is vitally important. The job is to analyze all deployed and planned US intercontinental ballistic missiles and their penetration aids (gadgets like decoys, chaff (thin wires of metal), and others) in terms of their ability to penetrate any Soviet anti-ballistic missile (ABM) defenses now and in the future. We have to devise a national program of R&D that will ensure the US missile force will be able to penetrate any future Soviet defense. And we need to be able to assure the President that our systems can penetrate our adversaries' missile defenses and that the US deterrent is real." Suddenly, I was thrust in the middle of real stuff about the real world. Ruina asked me to work with Lieutenant Colonel Ken Cooper of ARPA, a distinguished Army engineer, to draft a memo for me to bring to General Ferguson, Air Force chief of R&D. We worked out the road map of needed actions. (Cooper became a close associate, and eventually rose to be Chief of the Army Corps of Engineers. After his retirement he joined ITT briefly, where I had landed before him.)

I learned then that the real world runs on drafts, General Ferguson and I worked from drafts, and the agreement was a draft, to be "fixed up later." The new program went into high gear very fast, and I was able to brief the Joint Chiefs of Staff (JCS) in early 1962 about what had been done and where the program was headed. The briefing was held in the secure conference room of the JCS, the "tank," and was apparently a success. The Chiefs liked what they saw and endorsed our approach for the program. Additionally, I remember vividly having a peripheral argument with the famous General Curtis LeMay, then Chief of Staff of the Air Force. The argument was about whether the yield of nuclear weapons on missiles was more important than accuracy of the missile, and was somewhat secondary to the main topic. The

Figure 4.1. With the boys, Charley Chris, Tom and Paul, looking over the Pentagon, ca. 1962. CH personal collection.

question was which would assure more damage to the intended target, higher yields of nuclear weapons, or higher missile accuracy, combined with lower yield nuclear weapons. LeMay favored more yield, I argued for more accuracy and less yield. (The Chairman of the JCS, Gen. Max Taylor, friend of President Kennedy, took notes.) The argument went on longer than necessary but it was about the shape of the nuclear missile force, and had big implications about the future force. But I wanted to get back to the focus on our program. Yet all went well, and happily LeMay did not hold this against ARPA or me. Apparently the story about the argument was talked about outside of the meeting. (So much for security in the JCS "tank.") When people asked me how I had the nerve to take on the great LeMay about nuclear weapons, I reminded people that he and I had fired the same number of nuclear tipped missiles in anger, namely zero.

It was sobering to be dealing with the actual US missile force, and not just R&D. We set up major databases, very highly classified, that gave all the current operational data, and we analyzed how the US systems would work against a variety of Soviet defenses. It was a real Net Assessment, though the term had not been coined yet (the actual Office of Net Assessment was not established until 1973.) It was an exhilarating way to start, and a good way to begin to get to know the people with whom I would be working for many years. I also learned a key lesson: The main issue about ABM was not the details of the technical program, but how the program affected the strategic balance between the US and the Soviets. This "refocusing" helped my approach with all the other programs we had.

Project Defender

I met the key people in Defender: Al Rubinstein, who had run Defender until I showed up, and later became my effective and loyal deputy, for which I was truly grateful. Lt. Colonel Ken Cooper quickly became my right-hand man, highly regarded and highly effective. Others were Ralph Zirkind, the guru of infrared physics and engineering applications; Vic Kupelian, a great builder of very fast rockets; Jim Van Every, the king of high frequency radars (he had set up more of these large and complicated machines than anyone else in the world;) and many more. I met the JASON group, at that time quite secretive, which became an essential, and helpful part of my environment. The Jasons were a group of top-notch scientists who, in addition to their regular academic work, agreed to work part-time for the secretary of defense exclusively. They were very smart, and a real powerhouse of talent, very helpful if used thoughtfully, less so if handled routinely. And I was introduced to the Lincoln Laboratory of MIT, the premier scientific laboratory for all things radar.

I met "my" programs and the ARPA program managers and contractors. It was a huge program with an annual budget of some $120 million in the 1960s, equivalent to eight or ten times that in today's dollars. (To give some perspective, my NBS Heat Division budget had been some $3 million.) Defender had hundreds of programs. Some work was excellent, some middling, some was not sensible or important.

Figure 4.2. Ralph Zirkind at my 80th birthday celebration, 2005. Photo by Susanne Ollmann.

At that time the Defender program was deemed by many to be too diverse, with too little focus. I agreed, and quickly realized that it was a challenge to understand, let alone explain it. Therefore my top challenge was to figure out what was critical, what was less so, and what could and should be stopped. It was also important to evaluate why the program existed, what results could be expected, and in what time frame.

The big elephant in the room was the US Army Nike Zeus ABM program, which was under way to full development, and whose mission was to deploy a national ABM system as soon as practicable. It had a huge budget – several billion dollars a year.

The Defender program was intended to be the leading technical edge in this effort, but to stay independent of the Army program. Keeping it independent gave the secretary of defense enormous leverage over the Army system. This showed up in the half-yearly briefings I gave to the secretary of defense about our program. It was attended only by the secretary of defense, the Director of Defense Research and

Figure 4.3. With Vic Kupelian and Mrs Kupelian at presentation of medal. ARPA photo.

Figure 4.4 Hibex D4, a superfast interceptor, part of Kupelian's interceptor program. ARPA photo.

Engineering (Harold Brown), Jack Ruina, the Director of ARPA, but also the Secretary of the Army (several times Cy Vance) and the Army Chief of R&D. These were intended to be informal, "talk it over"-type of meetings, but occasionally the sparks flew.

- So my main tasks quickly became: To get on top of the Defender program, technically and managerially;
- To straighten out relations with the Army ABM program;
- To work out relations with the Air Force and Navy strategic missile programs;
- To craft a joint penetration aids program with them; and finally,
- To understand the impact of all this on the strategic balance, and learn how to explain it to the top policymakers, and the Congress.

The ARPA Defender program pioneered the most advanced technologies related to Ballistic Missile Defense, and did so nominally in support of the Army's Missile Defense Program.

Defender initiated or modified the whole gamut of missile defense technology. It pushed a completely new radar technology, including basic engineering science – the invention and development of phased array radars that could track many targets simultaneously, which transformed the radar field for good, and has found its way into almost all current uses of radar. For the radar and penetration aids work we had a large facility in the Pacific, on Kwajalein Island in the Marshall Islands. ARPA launched many diverse missiles from the Pacific Missile Range, and variety of missiles from other ranges at other targets, including from Cape Canaveral, and Woomera in Australia, and had several instrumentation sites all over the world, including radars on a large ship, to gather data from these launches. The program also sponsored the earliest US laser programs, when the laser power output was so low that it was just a lab toy, but that later were powerful enough to drill through steel and bring down airplanes and missiles.

We had a radar telescope in Arecibo, Puerto Rico, which was operated by Cornell University, which had a thousand-foot antenna fixed in the ground. It was a unique facility, whose use focused initially on measuring waves in the Earth's ionosphere, and evolved over time into doing wonderful research on planets, the sun, including radar maps of planets, until it was handed over to the National Science Foundation (NSF). It was a controversial program, many people recommended we stop it, but it was a great national facility, and helped cement relations with Cornell University, and with their great faculty in basic science. I am glad we did not get rid of it; it was one of the jewels in the necklace.

Signal processing turned out to be a technical key for almost every technology we were involved with, and while I knew little about it going in, I saw early on that it was a special application and extension of statistical mechanics, which I knew well, so I quickly became familiar with the basics. Lots of scientific things fell into place for me. My experience at NBS helped me clean up the uneven quality of the many

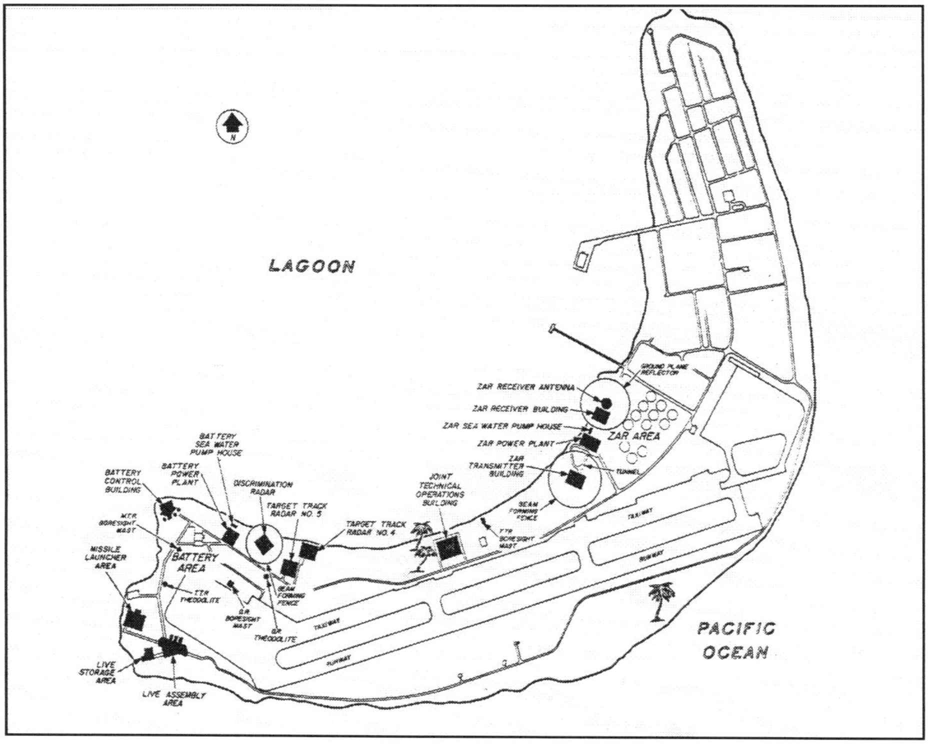

Figure 4.5. Kwajalein map, Photo by ARPA.

Figure 4.6. MIT Project Press at Kwajalein. Author second from right; AF Major General McCormick (retired) VP of MIT, third from right. Major Kiernan, US Army, second from left, Project Press Manager second from left, he was killed in Vietnam after his tour at ARPA, was commander of an Army Combat Engineer Battallion. Others are MIT staff. Photo by DARPA.

measurement programs. My theory background enabled me make ARPA analysis more rigorous.

As our analysis effort expanded, we began to understand the real arguments for missile defense. In those days, ABM systems were planned with interceptor missiles having nuclear warheads, and our nuclear weapons community was rife with advanced ideas for these. We had the best engineers in America to help us understand what was needed and how to accomplish this: experts and senior leaders from Bell Labs, Livermore, Rand and many more.

There was tremendous demand for important results – very accurate radars, and computers that could find small objects in very noisy data, very fast interceptor missiles and lots of field measurements on ballistic missile launches and missile interceptions. The Kwajalein facility played a special role. The Kwajalein Atoll had been occupied by the Japanese before World War II, and after the war had become a

Figure 4.7. Briefing at Project Press, ca. 1963. DARPA Photo.

mandate territory of the United States. The US Army had built a huge research facility there to test the Nike Zeus ABM system. ARPA had been assigned the island in the northeast corner of the beautiful atoll, called Roi-Namur, for our experiments. Roi, as we fondly called it, became so loaded with radar and optical equipment that we joked: "one more gadget and it will sink."

Well, we added more instruments, but sink it did not. The experiments we did there with help of Lincoln Lab, Westinghouse, RCA and others changed the way missile defense was understood, the results partly determined the evolution of the strategic balance, by showing what would and what would not work in the real world. The results were watched closely by the Washington policy community – and became a staple for my hearings before the Congressional committees. And they were watched by Soviet intelligence collection ships. We did, therefore some other experiments at the Woomera missile range in Australia, which were hidden from any spy ships. I went several times to Kwajalein and to Woomera, and got to know and respect the people there, and the difficult work being done.

Figure 4.8. Dinner in Australia, ca. 1966, at a meeting after a visit to Woomera. Herman Bondi (later Science Advisor, UK MOD), Tom Gold (astronomer from Cornell), Ed Salpeter (a Jason and later Nobel in physics), CH, and all born in Vienna. CH personal collection.

That all this was serious stuff become evident quickly, when Ruina took me to a meeting in the Situation Room at the White House, a fabled location in Washington. Back then, it was a simple conference room in the basement of the West Wing, and had no displays and no modern IT tech, just a few maps on the walls, some of them covered. (As I learned a little later, next to it was a very large room full of communications equipment and many human operators in real-time touch with the whole world.) The meeting was about the first real data regarding what looked like a potential Soviet ABM system. These data were presented by Bud Wheelon, the new CIA technical chief (who was to become a close colleague and friend), and the audience were "Mac" Bundy, President Kennedy's national security advisor, and a few other White House types. The meeting was serious and focused but relaxed, as became clear when Bundy asked who I was, and Ruina told him that I ran the ABM program and was supposed to know "all about it." Bundy jokingly replied, "Lord, who let him in?" These were very bright people with huge responsibilities, but they maintained a sense of humor and did not feel a need to hide it. The meeting ended inconclusively – too little could be inferred from the data, and not enough was yet known about the potential of ABM.

This basic uncertainty became characteristic of many of our endeavors in this and other fields.

I managed to persuade David Mann to join me from NBS, to help run the large and important re-entry physics program, and also persuaded Eileen Hansen of NBS to come to be my Executive Assistant, and help me run this very large, very complicated and very controversial program. This made for a much stronger team.

Why Do All This? Policy Help Arrives

I learned early that the hardest issues to understand and explain were what our technology had to do with the National Security Strategy, with arms control, and the strategic balance. The big national security

Figure 4.9. David Mann with his wife Vera, and Shannon, at my 80th birthday, 2005. Photo by Susanne Ollmann.

decisions were generally not made on technical grounds, but on strategic and political grounds. So I asked for expert help. In early 1962 I wrote three letters that turned out be milestones in this evolution: one to Henry Kissinger, then a professor at Harvard, who had written powerfully on nuclear strategy. I also wrote to Herman Kahn, who was about to leave the RAND Corporation to set up the Hudson Institute; and the third to economist Tom Schelling, also at Harvard. Each letter explained briefly what I was doing and why I hoped to get advice with the complicated strategic and policy issues behind the technical issues of ballistic missile defense. To my delight, each answered, differently but very constructively.

Kissinger invited me to give some seminars about ABM to his classes, which I did. This started a collaboration that lasted for two decades. Herman Kahn, who had written *Thinking about the Unthinkable,* (the policy challenges from the atomic bomb) came to talk to me about Hudson Institute. We wound up funding some of Hudson's earliest major studies, and a personal collaboration began that lasted until his death two decades later. Tom Schelling, later an economics Nobel Prize winner, famous for strategic studies, invited me to join a top secret strategic study he was starting in 1962. It was to make a ten-year projection of the US strategic posture, for the Departments of Defense and State. All three moves helped me greatly to understand the importance and limits of our technology, and "what all this technical stuff" was for, and it made ARPA and its output more comprehensible and helpful to policymakers.

Licklider, IPTO and the ARPANET

In 1962, a series of events began that changed a lot of lives, including mine, and arguably changed human history. JCR "Lick" Licklider, the real father of the Internet, came to ARPA. He gave several revolutionary seminars in the Pentagon about his radical ideas for making computing more useful. His talks changed my technical perspective deeply, and as it turned out they changed a lot of other things as well. He joined ARPA as head of the new Information Processing Technology Office (IPTO).

Lick's talks were the second life-changing infusion I received about computing, the first was by John von Neumann's lectures back in my graduate school days at the University of Chicago, and its impact on me is described in Chapter 3. Von Neumann explained how the first electronic computer, the ENIAC, had been built, how it worked, and that it was actually working on important and real problems. He projected a future in which electronic computers would take over many existing tasks, and make possible many new ones, ranging from pure mathematics to transportation, communications, and more. He predicted a radically different future, and he was right. The standard style of work in those early days was called batch processing, and was extremely awkward: You brought your "problem" to a "programmer" who took it, and after some months, handed you a pile of paper a foot deep, full of numbers that contained the "answers." You then had to dig through the pile to find what you were looking for. It was slow and unsatisfactory as a process for scientific research, but could solve problems that were totally impossible to do with a hand calculator.

Licklider proposed a new and totally different way, where the interaction between scientist and computer was direct, the interface natural, and preferably graphic. And these computers were all to be tied together in a country-wide net, so that people could use computers and data anywhere, and could interact easily at large distances. To some it had a ring of fantasy. Soon Lick started talking about the "galactic network," after all we were planning to put people on the moon, and these people should also be connected! But Lick was persuasive, and started the program that led more or less directly to the Internet that we have now. It was the genius of Ruina to see that this was worth trying, that Lick was the right man for the job, and that ARPA could and should invest in his idea.

As time went on, I spent a fair amount of time with Lick, in part because the ABM program used vast resources of computers but very inefficiently, and new approaches were desperately needed. Lick converted me to his broader vision. I become a disciple of his, and have remained so – in my office at home, there are various pictures of family, but only one of a colleague, Licklider.

In 1963 Jack Ruina left ARPA to go back to MIT. Who would take his place? Rumors suggested, correctly, that I was in the running, but Bob Sproull was chosen, and I became his deputy. (Needless to say, I was disappointed that I was not chosen as director, but got over it.) Bob was a very well known expert in solid state physics, from Cornell. He turned out to be a fine boss, and he taught me a lot. I was to be an "alter ego" deputy, i.e., in his absence I could act for him in all matters, and we divided the agency up for detailed supervision: He would focus on nuclear test detection, the materials science program, and the rocket fuel program, I became "godfather" of Defender, IPTO and AGILE. This was a lucky step for me, all three turned out to be successful, high profile programs.

Figure 4.10. JCR Licklider. Image courtesy MIT Museum.

IPTO again, and ARPANET

The main story of IPTO has been told by several authors and does not need repeating, but it may be useful to recapture my own reactions, thoughts and actions. IPTO was established to lead the Department of Defense's (DoD) R&D on Communications, Command and Control Systems, which were continually evolving – and necessarily, as it was obvious that the military systems were obsolete and in need of serious help. What was less obvious, and some have not yet learned it, was that these systems have to be in a continual evolving mode because the technology changes so fast, and the user's view of what they want to do changes even faster.

Working with IPTO was a great education in an exciting new field. We had collected some of the best minds in the world working on the program. In addition to networking challenges, there were fascinating programs on interfaces with computers, especially graphics interfaces, on the logic of computation, on the organization of information and its retrieval, and many more. One omission is exacting its vengeance now, however: We were not especially concerned about network security in 1965 and later, viewing it as something to be addressed "later."

A small, but important side-stream that drew much attention on Capitol Hill was game theory, which had just begun to infiltrate its thinking into strategic studies.

Information science was a new and fascinating world; most every stone that was turned over had a fine diamond stuck to its backside. Reporting about the program to Congress was actually a pleasure, because the high-quality and important work spoke for itself. And the DoD bosses did not interfere, as it was a relatively small program (less than 10 percent of the ARPA budget) and it also didn't bring negative headlines. So I almost always had time and money for the program; and later it produced the most spectacular results of any of our work.

The best example of such spectacular results, of course, is the ARPANET, that soon led directly to the Internet. The ARPANET was a long shot for the military systems, which was our official justification, because those military systems were a mess then. (Nowadays, almost all military systems use internet technology.) But the ARPANET and then the Internet, which was initially devised as a communications and computing medium for academics, and the World Wide Web, which was also designed for that purpose, met a deep need of every human being – a need for connection and expression, for human contact. Few of us had thought about or anticipated this – except perhaps Licklider, who understood it better than all of us.

Nuclear Test Detection

Another major ARPA program that was a major success also was related to arms control. This had to do with solving the question of how can one determine whether what looks like an earthquake was really a nuclear explosion underground? Rather than describing the program, I will use it to describe in some detail how it was done. ARPA was (and is) different from other funding agencies, and I was always unsure how to explain, when asked, but what does ARPA really do? So I will describe the "Art of 'Doing ARPA'" and will use the Nuclear Test Detection (NTD) program as the example. There are several reasons for picking this example: First, the program was very successful, and second, I contributed relatively little to its success, other than fighting for its budget. That makes it easier to be objectively descriptive about it.

The Art of "Doing ARPA"

There is much misunderstanding and confusion about what DARPA actually did and does now, and what its people actually do. How does one "do" DARPA? There is no training manual, no help desk, and little published so far is helpful. This note focuses on the actions required by all levels of the organization, not the regulations or the formal policy, or the "management handbook."

I believe that running anything is an art, a rather rare art, poorly understood and rarely described. This is also true for running a business, a department in a bureaucracy, or a school. Management usually describes the mechanics, the declaratory knowledge, so to say the "outside" of the activity, but not the implicit, non-declaratory knowledge coming mostly from experience, involving the "inside" of doing it. This thing has to be learned by doing, and it is best learned from masters of the art, it really cannot be learned from books. (This is actually a huge topic, which has interested me for many years. The implications of the main point are actually very wide, we focus here on one example, the "doing" of DARPA.)

I will try to describe the scope and range of things that are involved with doing a DARPA program, from the director's perspective, and I use a specific historical example, the setting up and execution of the Nuclear Test Detection Program of DARPA, as it was in its heyday, from the sixties through the '80s of the last century. (Choosing this avoids criticism of recent events, and it offers a view how somewhat quaint things used to be. I was a participant and actor at the start of this period, but not the prime actor. The example is valuable for showing the variety of factors that combined to make it work.)

Step 1. 1958, The President assigns the program to ARPA, the secretary of defense concurring. This is a critical step, without it many can challenge the program, and most who can, do so.

At this stage the program objectives are incompletely defined, but there is good agreement that some thing like it has to be done.

Step 2. The purpose of the program is more fully defined, namely to build a system that can reliably distinguish natural events, earthquakes, solar flares, storms, from nuclear explosions underground, in the atmosphere, in space and under water. The policy motivation for all is to see if we can identify nuclear explosions sufficiently reliably that then we can make arms control agreements to limit nuclear testing, with a formal verification and inspection regime, and be assured that if anyone cheats we will detect this cheating sufficiently reliably. Note that steps one and two are primarily about policy issues, not scientific or engineering or implementation issues, though the "reliably" is full of technical and policy puzzles, at least initially.

Step 3. Technical concepts are developed for implementing the Presidential directive.

This was done with involvement of several groups: ARPA, the President's Scientific Advisory Council (PSAC), DoD, Department of State, etc. We focus here on the detection of underground explosions, the other parts of the program exhibit similar features. In 1958 to 1959 ARPA had started, with White House and DoD urging, a large program of detecting underground events (explosions or earth quakes) using a large number (200 to 300) of seismic stations distributed over the whole world. By 1963 this program was running and showed promise. But more detailed detection was needed, especially the ability to distinguish earthquakes from nuclear explosions. The key technical idea to obtain this capability was to use a very large number (some 250) of sensitive seismometers, all tied together in a large array (spanning some 150 miles in Montana), and process the detected signals using beam forming algorithms, well developed by ARPA for radars in 1960 to 1961. This was a conceptual and technical breakthrough which I attribute largely to Jack Ruina, made over a time in 1962 to 1963. This was an idea of first importance, and a major technical discovery. The center of the array was in Angela, Montana, a "town" consisting of a general store, a Post Office, and a gas station.

The data were analyzed in a new facility in Billings, Montana. (It was likely that the concept would work, the laws of nature promised this, the discovery was that it most likely could be done, and that it "would change the game.")

Step 4. The next several items belong together, they all had to happen, they were all technical developments:

- improve sensors
- learn how to bury them in the ground for best effect

Figure 4.11. The town of Angela, Montana, the Post Office. Also a gas station, and farms miles away. CH personal collection.

- tie them together with special cables
- collect the signals and process them to get rid of noise,
- form beams from the signals coming from all directions, this required large new computers and computing techniques,
- learn to display the output of these computations so as to be useable, communicate and explain the results to non-specialists.

Note that these were all highly technical, challenging the state of the art, most had never been done before, and never had been put together into one system.

Step 5. The system had to be built and put in place. Much of this was done by the Air Force, but with ARPA help, and occasionally, when called for, with direct ARPA supervision.

Step 6. Next it was necessary to test the system. There were plenty of earthquakes around, and lots of Russian and US nuclear explosions. However, it was determined to be highly desirable to arrange several specially tailored nuclear explosions that could fill gaps in the calibration effort, and explore actual attempts at cheating measures. So ARPA arrange two specially designed underground explosions, carried out with the Atomic Energy Commission (AEC). Note that this step involved both technical and policy matters, experts from ARPA and others in the technical community. These data were analyzed and policy conclusions were derived. And ARPA people were heavily involved in all of that.

Step 7. The problem to discriminate natural events and manmade ones never went away. There was a constant debate, lasting for years, whether the system was good enough, and if not what to improve and how to pay for this. This was again a mixed technical and policy matter in which ARPA was involved directly. All these matters had to be planned and executed on a tight schedule and budget. The range of issues to be dealt with was enormous. The risks and the downsides were daunting, and the expense significant.

Doing all this and enjoying it was what we thought was the "doing ARPA," which I consider a high art form.

Project AGILE

When I became Deputy Director, I began to spend more time with Project AGILE and its chief, Bill Godel. I had met Godel during my first visits to ARPA, and had been impressed by him and his program. He was incisive, smart, and tough. He had served as a marine in World War II, had been taken a prisoner of war by the Japanese and held in a prison camp in Indonesia. As he told it, he got fed up with it and escaped ("walked out" is what he called it). He ran Project AGILE for a while, and after he was promoted to ARPA's Deputy Director for Management, he continued to be AGILE's godfather.

AGILE was set up very early in the life of ARPA and was designed to help the government of South Vietnam to counter the Communist insurgency raging there. It had a variety of programs and two labs, one in Saigon, South Vietnam, to help the South Vietnamese, and later one in Bangkok, Thailand, to assist this country with its growing insurgency threat.

My involvement with Southeast Asia deepened over the next several years and it became a major preoccupation, a source of great pride and also great grief. We were almost the only agency dedicated to help the local military directly, and that made our efforts different from all the others. It also gave us contacts with the South Vietnamese, and therefore more insight into what was really going on there.

Because of its international, political, and military scope, Project AGILE presented unique management challenges. I wanted to run all ARPA projects with a light hand, but AGILE was different, and required a more nuanced approach, given its complexities and risks. It was essentially impossible to make firm, precise commitments about the new ideas for work because we didn't know how much they would cost, how long they would take to explore, and whether or not they would succeed. All these problems are typical of any operations in developing countries, but that was not understood by most of our bosses. We wanted people who had experience, deep technical understanding, good sense and integrity. They also had to understand or learn the environment of Southeast Asia, as well as the policy objectives of the United States and the DoD. In addition, our local directors in Vietnam and Thailand had to be able to work with the local people, and be skilled in negotiating with them every day about every detail. Such people were very hard to find, but we had some stars.

Part of my effort to make the machine run well was to visit the overseas sites frequently and at length. As deputy, and later as director, I made usually three trips a year to all the major sites, each trip lasting about a month. They were strenuous trips – travel almost every night, up at dawn to visit the field or labs, meeting with the locals about their problems, various meetings with colleagues and local officials in the evening. But the lack of sleep was worth it; it was necessary to see the inner workings of the program in order to help everything run more smoothly. In these trips I preferred traveling east, it made for shorter days when traveling, and got me to Europe at the start. There I met with my UK colleagues, especially Jock Henderson, a colleague who was then chief scientist of the Royal Air Force, Herman Bondi, chief scientist of the MOD, and various other people in the British Ministry of Defense, (MOD). And I would try to stop on the way back for a weekend in Hawaii, where I wrote my trip report, and reported to the Commander-in-Chief, US Pacific Command (CINCPAC) and others what I had been doing in their area of operations. This last was most helpful, they learned from me some facts on the ground and they shared their perspective of the theatre command, including problems they were having with Washington.

More importantly, I believe you cannot send people into situations that are physically dangerous, as in Vietnam, unless you are willing to go there yourself. I had to go wherever my folks had to work. It was also important to show the local people that we, from the Pentagon, shared the risks they took. Being in combat was not for civilians, but taking substantial risks was important. I had learned in the Army that being shot at, even potentially, focuses your mind and sharpens your perspective.

Project AGILE: South Vietnam

Let us begin this story with South Vietnam, the original "front" of Project AGILE.

The war in Southeast Asia had been going on since the end of World War II. Vietnam, Laos and Cambodia had been French colonies before

World War II, and were then occupied by the Japanese. When the Japanese left, indigenous independence movements began, strongly influenced by leaders who had been trained in Moscow and were committed Communists. This movement was called the Viet Minh. They eventually threw out the French, and separate states were created by the United Nations. Eventually, Laos and Cambodia became independent (but not viable) countries. Vietnam was split. The north was dominated by the Viet Minh, who installed a classical Communist regime. Next, a million Vietnamese walked from the north, toward what was becoming an independent and relatively democratic state. Religious and ethnic diversity, and bitter political conflicts were deeply embedded in the area, and drove events in ways that were not always evident to the outsider. This was our battleground.

My first trip to Vietnam and Thailand occurred in late 1963, and it was another of those life-changing events for me. The trip was designed to take me to places where I could learn about counterinsurgency warfare: Japan, Taiwan, Malaya, Vietnam, and Thailand. On November 1, 1963 – my second day in Vietnam – President Diem of South Vietnam was killed in a South Vietnamese Army coup.

It was scary, but started routinely. The day after we, General Wienecke, then head of AGILE and I, arrived in Saigon, we took the ARPA plane, a deHaviland DH-7 (an experimental plane on loan to ARPA for testing,) to visit South Vietnamese troops around Can Tho, a town in the Mekong Delta, which was a center of insurgent fighting. We spent our time with the Vietnamese corps commander and his people, and we went to see some of the local units in the field. There were a few American Army advisors on site. Everything was new to me, and fascinating. At the end of the day, we flew back to Saigon. The most senior US official on our plane was Brigadier General Bob York, who was then the top technical advisor to what was called the Military Assistance Group Vietnam (MAG-V). Some of his officers came along, as well as a couple of ARPA staff. York was piloting. As we were approaching Saigon, York said there was trouble; we were told to circle,until eventually we were able to land in a remote part of the Saigon air field. Once on the ground we were told there was a coup underway; the South Vietnamese military was trying to overthrow President Diem, and it would be best for us to stay out of Saigon.

After a while York said, "Nuts to that, you guys, we are all going to my house in Saigon." So we went to his house in several vans and set up to be prepared for a siege. The military personnel had arms. (When I arrived in Vietnam, I was immediately asked, "Do you want to carry a weapon? You have been in the Army and have been trained." I said, "No, I'm not in uniform, so I shouldn't. But if I need one, please give me one." My status as "trained but unarmed" carried through my tour at ARPA.) We had shotguns and rifles in the house and the officers set up a perimeter; it was all very professional and they were in contact over the radio with American headquarters. There was a fair amount of shooting in the city; when you looked outside, the sky was full of tracer bullets.

After a while, York decided it was safe to send us to the Caravelle Hotel, in downtown Saigon. Major General Wienecke was with me for the whole trip. (He was a famous man in his own right. In World War II he was in the 82nd Airborne Division that went by air into France on D-Day, and he crash-landed in a glider in Normandy. He got a severe whiplash injury that troubled him for the rest of his life. He then retired from the Army but returned to run AGILE. He was a great fellow; I got to know him very well.) Bob and I went to the roof garden of the Caravelle Hotel to see the fighting in the city – heaviest at the presidential palace, at the barracks of the presidential guard battalion, and in a few other places. Everyone wanted to know who was winning, a more than casual concern. Once in a while a sweep of the machine gun tracers would come near us and we would duck behind a little building that was on the roof. Around 3:00 in the morning, I went to bed; there was still shooting, but not near the hotel.

A few hours later, Air Force Major Berkow, one of our program managers, picked me up in his car, telling me the fighting was over and it was safe to leave the hotel and drive around town. He was in uniform and carried a standard Air Force pistol, he took a revolver from the glove compartment and gave it to me, saying, "Here, put that on the floor between your feet, so no one can see it. If anybody pokes his head in the car, just shoot him, all right? You know how to do that?" I said, "Yes, I know how to do that." He said, "Okay, let's go."

He knew all the "good spots." South Vietnamese soldiers and tanks were everywhere. First, we went to the palace that had been taken by the coup soldiers. Crowds were milling around, waving happily to us Americans as if we had liberated them. We then went to the barracks of the palace guard where the fighting had been the heaviest. We got there just before they sealed the place. There was a lot of destruction and I saw several dead laid out on stretchers and covered with blankets. Then we went to the post office and a police station that had been burned; friendly crowds milled around both places. Later, at the harbor, we saw another crowd watching as a monument of the famous Trung sisters that had been set up by Diem – Vietnamese heroines who fought against the Chinese in the Middle Ages – was toppled.

Next we went to a briefing for General Harkins, the four-star general who was head of the US Military Assistance Group. He welcomed me, introduced his staff to me, and asked me to give him my reactions. As he got a run-down of the action at the places where we had been, I could whisper to him, "Yes, I was there an hour ago. That is exactly what is going on, and I saw all that." Then came the news that Diem had been killed. The official story was (and I have no reason to doubt it) that Diem, after being captured and kept in an armored personnel carrier, reached for a hand grenade of one of his captors, pulled the pin and killed himself and his brother, and several others.

And then I got yet another look at events. The American Ambassador, Henry Cabot Lodge, invited me to talk to him. He said he knew that I had been all over and he wanted to know what I had seen. And just at that time, he took a phone call from the widow of Diem's brother Nu, and assured her that she and her children would be safe and would be allowed to leave to go to Paris.

I'll never forget those twenty-four hours as long as I breathe.

While the rest of the trip in Vietnam was not quite so dramatic, I do want to give a sense of the scope of the projects ARPA worked on in the country. Our job at that time was primarily to help the Vietnamese people – not the US military directly. Project AGILE had a diversified program – hardware development, also analysis of Viet Cong tactics,

Figure 4.12 Marines moving into the Presidential Palace. Photo by CH.

Figure 4.13. Entrance to the guard compound. Photo by CH.

Figure 4.14. Inside compound where fighting was heavy. Photo by CH.

Figure 4.15. Where fighting was heaviest. Photo by CH.

Figure 4.16. Author with General Wienecke, third from left, in Vietnam. DARPA photo.

politics and economy – which resulted in a wide array of projects, some successful, and some less so.

One of our successful efforts was our work with the people who lived in the highlands. During my stay, I went up to the highlands to the Army Special Forces camp of Plei Mrong, where the US Army Special Forces were training local Montagnards for self defense. The Special Forces were quite different from the rest of the US Army, heavily influenced by para-military ideas and allegedly financed by the CIA in part. I went up to their camp, which was in the fighting zone, to see what the conditions were. The camp had been taken by the Viet Cong some months ago, retaken by the South Vietnamese, and was considered safe when I went there.

To get a social science take on these problems, ARPA had a famous anthropologist named Gerry Hickey working there in the highlands, who spoke some of the highland languages and also Vietnamese and had a deep understanding of the cultures. He described his insights in

Figure 4.17. Plei Mrong, troop training. Photo by CH.

Figure 4.18. Plei Mrong, training troops. Photo by CH.

Figure 4.19. Plei Mrong, people watching our departure. Photo by CH.

Figure 4.20. Gerry Hickey gets a medal from General Westmoreland, with John Foster, 1966. DARPA photo.

training manuals on how to work up in the highlands with the locals, as well as in scholarly publications.

It was clear that if we were going to be fighting that kind of war, we had to understand the local people. And in the highlands it was the highlanders; or Montagnards, in French. The Montagnards are neither Vietnamese nor Chinese; they have their own unique ethnicity and speak a variety of languages, which are dying out quickly. We studied the Montagnards to help the Special Forces to work closely with them. They had to understand the language, the culture, what do you eat, what do you not eat, how do you identify who the village chief is, how do you establish a good relationship with him, and so on.

In fact, the Montagnards were so pleased with American involvement that they fought with us well and some continued to fight even after the South Vietnamese had given up.

As we made our way around South Vietnam, I began to see the enormity of our task. We also quickly got "good news, and bad news." The good news was that the US advisors to the South Vietnamese (who were not working for ARPA) were settling in, some became quite helpful, others less so. At the time there were some 23,000 US advisors all over South Vietnam. The bad news we got was that President Kennedy had decided to reduce the number of advisors beginning Christmas 1963. (It is now a popular parlor game to guess "what would President Kennedy have done, had he lived." Well, in mid-November 1963, we thought we knew.)

This focus on the needs of the local military opened up a variety of other projects – the most famous of which was the development, testing, and eventual adoption by the US Army of the M16 rifle. It was the ARPA engineering effort that had the largest long-term impact on the Vietnamese and the American military.

The standard US infantry rifle in 1961 was the M14, a new and excellent follow-on to the revolutionary M1, the US World War II rifle (with which I trained during the Korean War). The M14 had been developed for war in Europe, and was designed to be useful for fighting over large distances, between 1,000 and 2,000 yards. It was heavy and long – as it turned out, far too long for the shorter Vietnamese soldier. In addition, most fighting in the jungles of Vietnam was at ranges of 50 to 100 feet, much shorter than that expected in Europe. So, by all measures of good design, the M14 was unsuitable for the Vietnamese conflict.

So, in 1961, ARPA had begun testing another rifle, called the "Armalite 15," designed by a brilliant rifle designer, Gene Stoner. ARPA purchased a thousand of these rifles, spare parts and ammunition, and equipped and trained a South Vietnamese Ranger battalion with them. This battalion was so successful in combat that much of the South Vietnamese Army and eventually the whole US Army was equipped with the new rifle, which, after some further development, was called the M16. And this is still the standard US rifle.

Ironically, the first customer ARPA had for the M16 was not the Army, but the US Air Force. General Curtis LeMay, Chief of Staff of the Air Force – the same man I argued with during my first briefing to the JCS about the Pen Aids program, bought 600 rifles for the security forces that he had guarding the Air Force missile silos. He didn't hold a grudge; in fact, he loved the M16.

Developing small arms ordnance is one of the most conservative professions in the world of weapon development. Rifles typically are 20 years in service; they are very hard to replace. I feel good that we had something to do with introducing a new weapon that is still, in its modified forms, the standard US Army rifle. So that was certainly the biggest, most visible and most important AGILE success.

But it was not the only one. Others were smaller and, though not necessarily hailed by US observers, made a difference to the locals. Boots, for example. We learned that American-designed boots hurt the feet of the Vietnamese soldiers. That was considered for a while as just Vietnamese complaining. ARPA did better. We hired people who were experts in modeling feet and discovered that the standard American boots really did not fit well to the feet of the people in Southeast Asia. So we developed tropical combat boots for the South Vietnamese Army. That is a special art because the boots get wet all the time, yet they must stay dry to be comfortable for the soldiers. Another unique task involved, of all things, a fish sauce the South Vietnamese are fond of, called *nuk mam*. Vietnamese soldiers did not go anywhere without it. It happens that *nuk mam* corrodes metal; they used to take it in tin cans but the cans got corroded and that spoiled the sauce. They tried

Coke bottles, but when the kids jumped out of helicopters, their glass bottles for *nuk mam* broke. Then they would sit down at the landing zone and scrape up the sauce off their clothes to have some with their food. Well, we found a material that was not corroded by *nuk mam*, and made bottles capable of holding it. It pleased the Vietnamese a lot, but it created very negative PR at home, with people mocking the high-tech ARPA for wasting time and taxpayer dollars dealing with fish sauce. But this showed a fundamental failure of those without experience with the culture. It was important for the soldiers to have their *nuk mam*. Why? Because they had to – there are a thousand such "oddities" in every culture. The lesson is to never laugh at something just because you don't understand it.

ARPA also worked on improving village protection. The Viet Cong would come in the middle of the night, abduct the chief of the village and kill him and his family to intimidate the population. It got so bad that a lot of village chiefs went to the town overnight, instead of spending the night in their villages. By keeping track of which village chiefs slept in town we got a rough idea which rural areas were controlled by the Viet Cong.

Figure 4.21. Visit by Col. Trach, Vietnamese army, head of our Saigon Lab, to visit my boss, at right side of Foster; John Foster, the DDR&E, Sy Deitchman, then head of AGILE, at extreme right. DARPA photo.

Figure 4.22. Col. Trach gave at talk at National War College. General Lampert, President of War College, at right. DARPA photo.

The main thing was to help the Vietnamese build a self-defense force. If the locals were well led and adequately equipped, they were able to fight against intruders. But many peasants were not willing to fight – they wanted to grow food, have babies, and lead a normal life. To ask them to fight required an act of commitment by and confidence in their government that most of them did not have, a commitment that they were often not willing to make.

To aid this effort, we were also able to help with sensors to give warning of intruders. We helped them make trip wires, and other methods. Later, we developed little radars that detected intruders. We also had night vision devices that people could use.

We tried out many things; some worked, some did not. For example, in 1965 it was awfully early in the art of night vision. The first generation device was a big tube, about half a foot long and weighing five pounds. It ran on batteries, and made all kinds of trouble – so that was not very popular, soon more advanced versions were developed and used; they became an essential component of American and allied night-fighting methods.

Bill Godel knew President Diem very well personally, so, as long as Diem was alive, ARPA had excellent connections with the South Vietnamese. But when Diem was killed in 1963, that connection broke down and we had to work with the people we found. And the quality of personal connections with top leaders is absolutely critical for success in such ventures.

As the war progressed it became clear that there was not that much of a threat from just the local Viet Cong. There were some people who were brutal, politically motivated, and were willing to die, but they were not numerous, and they were isolated. But it became clear that large amounts of weapons were smuggled into Vietnam by the North Vietnamese.

The Mekong River Delta is one of the great river deltas in the world, with a tremendous amount of traffic. There were occasional freighters, but most of the traffic was local, mostly fishing or transport. A lot of rice came down into South Vietnam from the highlands in boats that were specially built for carrying rice and that were floated down, and then towed up when they were empty. There were local junks all over.

The Americans and the South Vietnamese established some airborne and seaborne patrols to keep checking the traffic for contraband. Somebody in AGILE had the good idea that maybe you can tell where a boat was from by the way it is constructed. If you had that figured out, you could begin to understand the transport system of the adversaries. It turned out that this idea was exactly right. Our goal became to make it easy for the airmen and seamen to decide whether a junk was possibly hostile and worth boarding. So we had contractors take the measurements, take pictures, make drawings, of how junks from different areas of Vietnam looked – they were real works of art. And we published these elaborate compendia and technical detail of all the types of junks and where they had been built, a sort of "All the South China Sea Junks", similar to Jane's *All the World Warships*, one my favorite books when I was young, and is still. They were used in many of the patrol planes and ships of the US and Vietnamese forces. My hope was to do this analysis for the whole South China Sea; I expected violent wars throughout the region. The collection was never completed.

Figure 4.23 author on river Mekong, at helm of early development of an "armed junk". DARPA photo.

1965, The Turning Point in Vietnam

The turning point in Vietnam came in the late spring of 1965, and it was triggered by increased infiltration of North Vietnamese regular soldiers in large units. They came in small numbers, initially, but eventually a whole regiment came and then a division, and another division, until the final buildup at the peak of the war in 1970 or so. There were several regular Army divisions with tanks, artillery and full support from the North, which had captured large areas and had made some areas totally inaccessible. In 1965, President Johnson decided to send American ground troops in real numbers and more US Air Force and Navy activity.

The enemy buildup was gradual, and, as in most such situations, the reaction by the US was impeded by "not wishing to provoke the enemy," which is usually a hopeless quest. The North Vietnamese were doing what they were doing because they wanted to do it, not because of provocation by us. (This is a recurring theme. It does not mean you should not try diplomacy; it means do not bank everything on that. And for God's sake, do not worry too much about annoying people who are totally hostile already.)

The Tonkin Gulf incident happened about that time, and the serious bombing of North Vietnam started as well. It was a major escalation because we were clearly not doing well in South Vietnam in a number of ways, and it changed the war a lot. General Westmoreland took over and US Military Assistance Group (USMAG) was transformed from military assistance command to a military Army command, Military Assistance Command, Vietnam (MACV). It developed a life of its own that ran parallel to the counterinsurgency war, and overshadowed it more and more. As my visits continued, I got to know Westmoreland well, he invited me to stay at his house in Saigon whenever I was there, and we had lots of chances to discuss all these issues. His military strategy struck me as sound (he dispersed any large accumulations of North Vietnamese troops before they could do serious damage), but, in my view, he neglected the threat from the insurgents, who ate away the planks he was standing on. He was a man of tremendous integrity and I liked and respected him. He sent me a thoughtful and kind letter after I left ARPA.

Figure 4.24. Author and and Wienecke travel again (it is hot). US Army, 1963. CH personal collection.

ARPA got more involved at this escalation. We did a number of new things. As usual, some of them worked; some did not. And I traveled more locally.

Advanced Sensors Go to Vietnam

The advanced sensor group in ARPA, a new office that I had started, for some special projects, began helping the US military in Vietnam. There was a bridge in Vietnam that our military had tried – unsuccessfully – to destroy by air. Many enemy troops came into South Vietnam over this bridge, and many US airmen died trying to demolish it. We thought we would try to build some laser-guided bombs that would follow a laser spot that was put on the target. That laser could be either on the same airplane that carried the bomb or a different plane, or on the ground.

Well, we built eight bombs like that. They flew in 1967 and all eight failed. Some went into the water and did not explode, some flew between the bridge girders and did not explode. In any case it was a total flub. After I left ARPA, the program fortunately continued, they built a second round of bombs, and a year later finally dropped the bridge. I believe that this is the first clear step to what became precision warfare, where we used high technology directly on a weapon to make it hit the target. It became the hall-mark of American war fighting ways, lasting into today's conflicts.

When the decision was made to build "a barrier" to infiltration from the North, the sensors we developed were put together in novel ways, but not by us, another organization was created to do that.

Lessons Learned

One of the main lessons I took away from the Vietnam experience was that there *are* effective and ineffective ways to fight a counterinsurgency war. For the most part, counterinsurgency is not about military problems, it is about politics and power. It is about who keeps the people safe, who feeds them, who protects them. And he who does that will almost always win in the long run.

Another conclusion was that Asia is really different – it is completely unlike Europe or North America, from the food to the culture to the rules. (This is not news now, in 2013, but is was very much news in 1966 or 1967.) Asians believe large changes take time, so they are suspicious of instant solutions. I would say that I also have also become very suspicious of instant solutions. Big changes take years; it is easier to claim success quickly and walk away from the problem. The notion that one can waltz through a large complex problem and have a quick "transformation" is fiction. But it is the favored conventional approach to today's problems.

So, for me, this first trip to Vietnam led to a multi-faceted voyage of discovery. I'm proud to have been able to participate, and this feeling is with me still.

Project AGILE: Thailand

Thailand was the second “front” for Project AGILE, and it was the second part of my first trip to South East Asia. In early 1961, ARPA had made an agreement with the Thai government, much like the one with the South Vietnamese, to set up a joint R&D laboratory near Bangkok. It was staffed the similar way, with a Thai commander (Air Vice Marshall Manob) and an American number two, with ARPA paying for most of it. The Thai contributed many good people.

After my 1963 trip to Vietnam, I went to Thailand next. General Wienecke accompanied me and we met a lot of high-level people, including Air Chief Marshall Dawee, who was then head of the Thai Air Force and then later became head of the military and eventually Prime Minister. In addition, I saw the ARPA lab, met many of the people, and saw the projects they were working on. At that time, there was no very clear focus on what to do.

Figure 4.25. Thailand, with troops. CH second from right, John Phelps, my executive assistant, at left. DARPA photo.

Figure 4.26. Vice Marshall Manob, right, head of the Joint Thai ARPA Lab in Bangkok, and General Stilwell, US Army, 1965. DARPA photo.

Figure 4.27. Marshall Dawee gives lunch for author, 28 Nov 1966. Air Vice Marshall Manob, Thai Director of Joint Lab, left, then US Ambassador to Thailand, Mr. Graham Martin, Air Chief Marshall Dawee Chullasapya, the author, Mr. de Silva from US Embassy, Lt General Leg, Thai Army, Richard Holbrook, US Director of Joint Lab in Thailand, on right. DARPA photo.

What was immediately clear to me was that Thailand was special. It was new to me, and I found the people charming. When you walked in Bangkok, people smiled at you. The Thai tend to be very friendly people. Thai art, which mostly revolves around the Buddhist temples, was extraordinary. I spent a fair amount of time wandering around in temples; in those days, you could just take your shoes off, go in and sit on the floor, and just look and listen and do nothing. No one was bothered by that; it was perfectly appropriate behavior. It gave me a lot of time to reflect – which was good, because there was a great deal for me to process.

What became obvious, as we understood better what was going on in Vietnam and in Thailand, was that the Thai insurgency was in a very early stage, and mostly local. Our idea and hope was to try things against the early stage insurgency and have more time to try different things if the first experiments did not work. And that was what happened. We tried a number of things, including some really dumb ideas. For example, many villages had a single centrally located well, and all the women went to the well and then took the water home. We thought it would be a great idea to just lay a water pipeline from the well to every house.

So we did that and found out very quickly that the women still went to the well to get the water – because that was the social center of the village. And if they wanted to hear about "Aunt Betsy," they had to go to the well and ask Aunt Betsy's daughter. All the socializing took place at the well, not in the home, and our Western ways were not attuned to their customs. It was a small but stunning hiccup, and instructive in many ways.

Thailand also was a different challenge than Vietnam, partly because the Thai had a very clear idea that the military were the local presence of the government, and they would do everything necessary to prevail. There was no hesitation at all about helping the population with food crops, fishing, medical problems, and so on. But when we showed up in order to assist, they did not really know what they wanted help with. And if your customers do not know what they want, you have a tough time helping them.

I returned to the United States around November 20, 1963, and made a stop in Milwaukee to give a talk about my trip to the Far East. On November 22, I was waiting in the airport for my plane back to Washington, when the news came of the shooting of President Kennedy in Dallas. I was devastated. I had been an enthusiastic member of his team, one of McNamara's "whiz kids." In some sense it was the end of an older world, and when I got back to the Pentagon I found that everybody I knew had the same feeling of confusion and abandonment, and a great reluctance to grapple with the "Vietnam problem."

Yet the problem had not gone away, and there was still work to do, so on we went.

Later, the Thai got nervous about the northeast and said, "We have to do something there," and then we began to see an opening. Project AGILE was run at that time by Sey Deitchman, a great "ideas man" and good manager. We had a very good guy in Bangkok at that time, Dick Holbrook, who was a fine engineer and a good manager. The three of us decided to do whatever we could think of to help the Thai protect a single important province in the northeast, and asked the Thai to choose the province. They selected the Nakom Phanom Province, on the Mekong river across from Laos.

I traveled to the province several times, and got to know the area and its province governor, a serious guy who got things done. I found that it always matters if the local people in charge whom you would like to help, trust you. If they do have confidence in you, you must never disappoint them. It's a careful balance between doing what they want you to do, and doing what you want as well.

In Nakom Phanom Province, the Thai wanted a defensive surveillance line along the Mekong, which is the border between Laos and Thailand, where they could detect and stop infiltration from Laos by the insurgents from the north (the North Vietnamese and, possibly, Chinese.) And so we said, "Okay, we know how to do that." First, we showed them a variety of sensors, radar, IR, good telescopes, and cameras, and said, "These are some of the things we can put together, but we suggest you do a system that is integrated." They agreed.

We also said, “We would really like to work together. You will have to do most of the work but we will provide help.” Together, we produced a great planning tool – a detailed photo album of both sides of the Mekong from the north of the province to the south of the province. It took months, and was very detailed, it covered every stair from the banks down to the water, every little dock and every little cave, in high-resolution.

This detailed analysis of the Mekong shores was worthwhile, producing all kinds of interesting data. We studied how many boats the average family had depending whether they live on the Mekong or further inland, and discovered that the average family living on the Mekong had about three boats (it was an exponential decrease as you went inland). More importantly, the data suggested where to put radars, infrared and acoustic sensors to get maximum protection for the area.

We also found that there was an enormous amount of cross-marrying across the Mekong – Laotian men marrying Thai women and vice versa. People would settle on either side, so there was an enormous amount of innocent traffic. The challenge was how to find the bad guys – and there, the local police were the experts, and effective, because they knew everybody.

In the end, we put together a comprehensive system, and it was well-run by the Thai police and military. There was at that time a significant US military presence in Northeast Thailand, principally US Air Force bases, and when the “wall of sensors” was set up in Vietnam along the Ho Chi Minh trail, the analysis and processing center was put in at Nakom Phanom, as a central post which got all the data from the sensors.

In 1972, I had the chance to visit that center. It was a fusion center for all information in the area, including the data from the field of sensors. And I could hear Russian trucks on the microphones on the Ho Chi Minh trail. They had a very distinct audio signature when they were shifting gears. The contacts were reported to the Air Force, but no action was taken.

A final word about the province governor: He once took me on a boat ride around on the Mekong in his "yacht," a medium sized motor boat, we were accompanied by two or three police boats that came along lest somebody try and whack him. And I asked him, "Are you worried about your safety?" He said, "Yeah, I'm worried. But I'm not very worried." And he took me down below deck and showed me a sort of marshal's baton that he carried, which had a hidden pistol in it. It had one bullet in it. Well, eventually word got out, and they sent two guys after him. He shot one, but the other shot and killed him – a reminder that this was a real war.

But it was a different kind of war than in Vietnam. The Thai succeeded, and that has made all the difference. The Thai were not going to be messed with – they were tough about their national independence and the integrity of their government. It was essential for us to understand early on that the Thai are enormously proud of their tradition of independence. They are very independent-minded. They are very tough. They can be very, very difficult to deal with. Unless you do it their way, nothing happens. So pretty soon, you realize you have to do it their way, and then say, "I would very much like to help you. Can you tell me what you would like me to do?"

I was very clear about that in all my dealings. The most senior person I dealt with was the Air Chief Marshall Dawee, whom I got to know quite well and who was close to the Thai King. The idea that you had to work with the Thai, and the challenges that could produce, are well described by one of the funnier stories of my time in Thailand.

In Thailand, nothing could be done without the King's explicit permission; One day, we got a frantic message in Washington from Bangkok. At the time we were looking at micro airplanes that the Swiss were making and we planned to use in Vietnam. The Swiss were selling kits to assemble for, as I recall, about ten thousand dollars a shot. The "crisis" was that the King wanted one of these. The King, in those days did a lot of hobby work, and he wanted to build one and to fly it. So we said, "Oh my God." First of all, he was insistent – and we knew that if we were to be allowed to have any sort of a program in Thailand, we would have to make it happen. But the US Embassy

Figure 4.28. Foster visits Thailand in 1966, with Sullivan, his deputy, 3rd from left, Dick Holbrook, right, Manob, center front, Col. Berkow, 2nd from right. DARPA photo.

in Bangkok worried, the State Department worried, the DoD worried, and of course, all the lawyers worried. And most of all, I worried.

We had two problems. One was: Could we do this legally? Thankfully, a bright ARPA program manager thought, "Let's give them two kits for training at the Thai laboratory, and the lab can do the training. We had bought the Vietnamese the AR15/M16s. We can do anything reasonable for proper training." So we did that and that worked.

But the second problem was, what if the King has an accident? So our task was to make sure that the King would not have an accident. Thai Air Force officers were assigned to help the King with this project. They did their job: all went well. This was just another typical day for ARPA management.

Becoming Director

After two very productive years at ARPA, Bob Sproull left to resume his academic career at Cornell. I was appointed director in 1965. Bob Frosch, who before ran VELA (the nuclear test detection program), became deputy director. We quickly set up a system similar to the one I had with Sproull. Bob was an "alter ego deputy," and we divided the programs between us for "god-fathering." Bob was the best deputy I ever had. He had a deep knowledge about a wide range of science. He had great good sense, and wrote exquisitely well. Besides, he was, and still is, a great person. The year we worked together brought a warm glow of professionalism and teamwork that has never been equaled for me.

But change was constant. Within a week of my taking on the directorship, Harold Brown left as (DDR&E) to become Secretary of the Air Force, and life changed some more. Johnny Foster became DDR&E. I knew him well from many interactions in important study projects like StratX, about decoys, and MRVs for strategic missiles. These had defined the scope and directions of future US strategic weapon systems, we had both been active contributors to them. I thought that he was very good and we hit it off right away; however, problems eventually developed and became serious later on. Becoming director provided an opportunity for me to fix some problems that had developed. In particular I refocused some of the projects, and that became my prime activity.

The most important of these new efforts affected IPTO. The program had reached a certain level of success, and Ivan Sutherland, its director had left. While good work had been done, I did not want to do just more of same. So after lots of thinking and consultation I concluded that we should get a, new, clear, and ambitious goal: Tie together all the good gadgets and techniques and see if we can build a working network with these pieces.

We hired Bob Taylor as new director for IPTO with the explicit goal of trying seriously to build a network of computers and associated communication devices. No one had tried this on a large enough scale, using packet switching. (The telephone system used line switching, an old and much less effective method, and the phone companies disparaged and fought the new program.) It was far from obvious that it could be done, and many voices weighed in with opposition and doubt. However, the idea was attractive and plausible, and we freed up resources for this. And if it was too hard, I preferred that we should learn this sooner rather than later.

So we proceeded with this. The details have been well told and do not need repeating. The project was a success. ARPANET worked, and showed it could be done. After I left, ARPA continued the stream of work and, later built the Internet, based on the critically important Internet protocols, "IP" and "TCP", developed by Bob Kahn and Vince Cerf, both at ARPA after my time. And that was a project that, over time, changed the world. But the outside world also changed, and we adapted to this.

Project AGILE: Lebanon and Iran

Project AGILE had two smaller programs in Lebanon and Iran, in addition to Vietnam and Thailand, and these now took more of my time and energy. The Lebanese office, opened around 1965, usually had a senior military officer and one or two other employees. It was intended to be a window to the Middle Eastern world, and was wonderful because our people there had good contacts with everybody in the area. In those, more peaceful days, you could travel almost everywhere in the region. There were minor conflicts, but nothing too serious.

Foreseeing that the Middle East would be a major future national security challenge – and that understanding the Arab world would be critical – we contracted with the American University of Beirut (AUB) to do a historical study of the most important tribes and families in the Arab world, who married whom, who was whose child, who was whose grandchild. The motivation was simple. There is a hidden history in the Arab world of family relations, this is true everywhere, but especially so in the Middle East. If your grandfather insulted my grandfather, you and I cannot be friends. It is up to me – my honor is at stake – to get the insult cancelled. The idea was to try and uncover who were the powers, who hated whom, who was friends with whom.

This AUB program was intended as part of a general Arab history project, stressing cultural and political matters. But when the civil war in Lebanon started in the 1970s, that was all destroyed, and some of the university people doing the work were killed. It was a sad story.

In Iran, the Persian part of the Middle East, we also had a small office in Tehran. Iran, at this time, was a firm ally of the US, and the US Army was the principal helper of the Iranian Army. ARPA was set up to work with the Iranian Gendarmerie, a large organization with very high-level political connections. A US Army colonel, who worked out of the US Embassy, was the ARPA person in Iran. We did a number of things with the Gendarmerie, including communications. The Iranian Gendarmerie had terribly inadequate communication systems, so we got them some decent high frequency radios and transmitters. Many of their posts were in the middle of the country where, there were

Figure 4.29. Col Tabor, our man in Beirut. DARPA photo.

no telephone lines. To demonstrate the dearth of communications, they took me to a town called Nain, which is in Central Iran. Among other things it is famous for the exquisite carpets made there. They took me there in small airplanes, and I was in the third airplane to ever land there, on a gravel road, because there was no airport. We went in two-seater planes; the one I flew in was piloted by an Iranian second lieutenant who learned flying at Fort Rucker, trained by the US Army – an excellent pilot, nice guy, very smart.

He landed on the gravel road comfortably, and then they took me around by car. I saw there were no communications to the outside. They had mail by truck, and a single high frequency radio, which did not really work. I was readily persuaded that they needed help, and we focused on that. And I saw how they made their fabulous rugs.

The other big thing I remember vividly is an effort to help the Gendarmerie with controlling the flow of drugs. There was a vibrant market for opium – something I saw up close. After an all day meeting in Tehran, where we talked about interdicting the opium traffic, I said casually, "You know, I have never seen raw opium." I had hypothesized that some of the solvents that go into the processing have distinct molecular signatures that a small mass spectrometer should be able to detect. This turned out to be true.

That evening, when I arrived at the colonel's home, there in the foyer, in the silver plate where people leave their visiting cards, was a brown brick of raw opium, not wrapped – a kilogram. It was just sitting there. My host looked at it and said, "Wonder what this is?" His aide replied, "This is opium, sir. Dr. Herzfeld said he wanted to see some, so we brought some."

So I smelled it, and it had very distinctive odor. I was satisfied that was worth pursuing, and we did.

The other thing we focused on was the opium smuggling inside and out of the country. Usually, this was done in water trucks. Water is a rare commodity, and usually was taken from Point A to Point B in big trucks, like our fuel trucks, with a top cover you could take off. Smugglers put three compartments into the trucks, and filled the center part with water and both ends with contraband opium. Ralph Zirkind quickly realized that infrared cameras would show it every time, because the different compartments would have different temperatures. And so we were able to find the stuff. But after a while we were told to stop these efforts, because we had gotten too close to the top level traffickers. That ended that.

I traveled a lot in Iran, and it was a fascinating place. Iranians are a very serious, proud people with a long history of great deeds and great thoughts. Some of the best Muslim literature is Iranian. Omar Khayyam, the great poet of the Muslim world, was Iranian. Their old mosques are some of the most beautiful buildings I have ever seen. I was truly pleased to be able to help, and to learn about their country – and I'm very sorry about all the problems they have now.

One of the lessons I learned from that time was that our tendency to be totally America-centric in our outlook limits the number of people we can work with. I think that is changing, however. I know lots of people now who are in American corporations, or in the US government, but work overseas, who are perfectly comfortable in other countries. They know the locals. They understand the culture. They are much less narrow in outlook, and realize that American ways are not always the best or the only good ones. I think we are growing up in that sense. Our military is now very different in this respect, and Iraq and Afghanistan have tilted the balance, again, in favor of people who are interested in the local scene. Whether that will affect the military for the long term remains to be seen.

Project Remote Area Conflict: India

India was another place where ARPA developed a presence, and another place that became quite special to me personally. It all started in October 1962, when the Chinese attacked India, both in the west, in the Kashmir areas of the Himalayas and in the east near the Indian province of Assam. The Chinese attacked ostensibly over border disputes. There was fighting at very high altitudes, from about 12,000 feet and up, as high as 18,000 feet. The Chinese had had troops up there for a long time; therefore, they were acclimatized to the high altitude. The Indians rushed troops up from the tropical lowlands, and were totally unprepared and poorly trained. The Indian leadership did not think they needed altitude-acclimatized troops, which resulted in disaster, and the loss of substantial territory to China.

The State Department and the Office of the Secretary of Defense asked ARPA, "Is there anything we can do for the Indians?" And there began an effort to see if the Indian military needed help, and specifically, if the Indian R&D people needed any help. And the inevitable counterparts to that question: "What can we get in return that we would like? What do they know we do not?" This was dropped into ARPA's lap in 1963. I was still deputy director at the time, and I was asked to figure this out.

The obvious first item of interest to the US was high-altitude medicine and operational experience at high altitudes. We were put in touch with the Indian Embassy in Washington. We began meetings with their defense attaché and we told him that our R&D folks wanted to help, starting with high-altitude medicine. It was clear that, first of all, they were really interested in getting help. Second, there was a good solid tradition

Figure 4.30. Dr Bhagavantam in my office at the Pentagon, after several years of investigating joint US Indian R&D programs. The pictures on the wall are from my visit to Leh. CH personal collection.

of people who were glad this cooperation would be possible and who had a bigger vision than just high-altitude medicine.

So we all said, "Let's explore," and thus began for me a number of unforgettable trips to India, to see where our DoD could help. I traveled as guest of the Indian Ministry of Defense (MOD) and of the science adviser of the MOD. And there I found a pleasant surprise, for the science advisor was a physicist called Bhagavantam, whose name and work I knew. He knew my scientific work and we were in similar areas, a field of mathematics called group theory (mathematical theory of symmetry) applied to properties of materials. He wrote a very nice book, which I had read and liked, and which I still have (which pleased him) and he had read something on symmetry I had written. So it was a good start; one cannot do better than that.

We became very friendly and he became a mentor, my guide to India. The deputy science adviser to their MOD was an applied physicist by the name of Ranganathan, who had an operations research background and oversaw a lot of the operations analysis in India. He was the person who took me around India and was with me almost all the time I was there. We also became good friends.

From 1964 to 1967, I looked at about twenty different Indian military R&D laboratories and a couple of non-MOD labs, and formed sufficiently clear impressions about their needs and capabilities. They ranged from the high-altitude studies, to electronics, to ordnance, and other fields.

My first trip in India took me up into the Himalayas, up into the Leh Valley, which is very high – the bottom of the valley is 10,000 to 12,000 feet. There was an easy way to go from Delhi to Leh, to take an Indian Air Force plane made in Russia, an An-12, that went regularly directly from Delhi military airport to Leh airport. But they started apologizing: "We cannot take you that way so we will have to take you some other way." When I asked why, they said, "It is an An-12. We promised the Russians not to let any American see it." I protested, telling them I knew as much about an An-12 as anyone would want to know, but they were firm: "We have promised them. We have to keep our promises." I said, "Okay. So how do we go?" That's where things got rather interesting.

Figure 4.31. Briefing in Gulmarg CH third from left, Dr. Ranganathan, fourth from left, Major Talwar, who traveled with me a lot, at right. CH personal collection.

First of all, we flew by Air India, in an old Indian Airways DC-3, to Srinagar, in the beautiful Vale of Kashmir. When we left Delhi it was hot and I was wearing a summer suit. When I got to Srinagar, the hosts said if we hurry we can get you by car to see Gulmarg before it gets dark, that way you will see our Army high-altitude training camp, an opportunity not to be missed. So we did all that. I caught a terrible cold, because Srinagar is at 6,000 feet and was quite cold. We drove up to 8,000 feet in an open Jeep and arrived just before nightfall at the high-altitude training camp. It was interesting to see the Indian Army cope with their many challenges.

The next day we took the Indian Air Force supply line over the Himalayas from Srinagar to Leh. In those days, about twenty airplanes flew every morning to take supplies up because the only way to get there was either by these airplanes or by a dangerous, very long, snowy truck road. The airplanes they used were old American Flying Boxcars, World War II vintage C-117s, with two turboprop engines on two tail booms and the big box for the cargo, and cockpit. They had to get through a high pass in the mountains – about 18,000 feet at the bottom of the pass. Because of the way the planes were loaded, they had a ceiling of 19,000 feet, so they just barely got through the pass. If there were clouds in the pass they could not go. (This was long before reliable navigation systems.) So one airplane took off first to see whether the pass was clear and then it radioed back "all clear," and everybody else would take off.

The morning we left, they sent the scouting plane out and got the other twenty planes ready on the runway for takeoff. They said, "Come on, you are the boss. You come in the first airplane. And if you want to see something, you do not have to sit in the cargo bay. Come and stand in the cockpit behind the co-pilot." So the engineer and I stood together, behind the co-pilot, neither of us having a proper seat, nor an individual air supply. We shared a single air tube. As this old cargo plane roared down the runway, the number two engine (the right one) suddenly quit, just before we lifted off.

The pilot and co-pilot were Indian Air Force pilots; they must have been in their twenties. These kids wrestled the plane to a stop and said, "Okay, let's go to the end of the line and see if we can get it to work." By the time we made it back to the end of the line, they had gotten the engine to work properly and they were sure it was fine.

So we took off and we made it. The altitude was a challenge – the only people in the plane who had oxygen masks were the pilot and co-pilot. (A few passengers in the cargo hold had nothing.) So the engineer and I passed the air tube back and forth. I was young and I thought it was unbelievably exciting and fun. It was incredibly beautiful to fly through the very high, snow covered mountains on a clear day, sun shining brightly, snow all over the peaks.

To get into Leh airfield you have to make a sharp bank into the Indus valley, which is there about 100 miles long. You have to bank sharply to get lined up with the runway and just left of you is big hill, maybe a couple of thousand feet higher than the bottom of the valley. As the pilot made the approach, he said, "You see that hill there? We lost a plane there last winter. Everybody died. It took us three months before we could get the bodies out because of the snow." Thankfully, this was in October, so it was dry and we landed fine.

Figure 4.32. In the Himalayas, looking North-East, toward Everest, which is far away. Photo by CH.

Figure 4.33. Another part of the mountains, looking East. Photo by CH.

Figure 4.34. A stupa, a shrine built by travelers, one stone at a time, at the point where the road leaves Leh, and goes down to the Indus river. Photo by CH.

Figure 4.35. rare traffic on road from Leh. Photo by CH.

Once up in Leh, they showed me a lot of the things that they were doing – which basically amounted to digging in, because this was disputed territory. Experts explained to me what was going on. They were building roads. They were building a larger airport. They put up communications. The agreement we had with the Indians was to keep me out of the tactical military stuff and just to focus on R&D matters. But when I came back, my sense was the Indians were digging in to stay. And all the events in the last forty-plus years agree with that.

The countryside was unbelievable. I traveled in an open Jeep again, but this time had a heavy Indian Army animal fur coat, leather on the outside, hair on the inside. We went from west to east along the Indus Valley, along the Indus River. The road was unremarkable, hour after hour, mile after mile, the scenery looked exactly the same. You had to really pay attention to the land on the sides to see whether you had made progress. And when you do that, you realize how big the land is. I have one picture looking far ahead, where a little patch of green

Figure 4.36. On the road from Leh, going East, note the oasis, seemingly near, but far away. Photo by CH.

is visible. I asked, "Grass?" They said, "No, no. Wait." It was, in fact, a huge oasis with tall poplar trees that had looked like grass from far away. The perspective was wild; it looked like grass miles away, but was really trees, miles away.

During this trip they also took me to a Buddhist monastery. The temple inside it had remarkable art – very dark, but colorful – not peaceful, but menacing. And lots of ghee, melted butter, burning in brass vessels with its distinctive smell.

An interesting coincidence happened while I was up in Leh: the first Chinese nuclear test explosion took place. Everyone had expected it to happen soon, but it was still a surprise when it did. The day before the test we were having an outdoor lunch, hosted by the Local Indian commander, who was a brigadier general. We were discussing military high-altitude problems. And then he said, "We hear rumors about the Chinese bomb. And you chaps keep saying it is going to be a small bomb. How do you know? Maybe it is going to be a big bomb." I replied, "No, it

Figure 4.37. A Buddhist monastery. Photo by CH.

is going to be a small bomb. I'm not allowed to tell you how I know, but we are sure." The very next day, there was a small bomb explosion and the quality of American intelligence was praised. ("small" at that time meant fission only, like the Hiroshima bomb, and "large" meant a fusion bomb, with a much larger explosive yield.)

My main work started after this memorable beginning. Ranganathan and I went to visit the laboratories and spent a day or half a day or so in each. They would tell us what they did and we discussed whether we could help. The subject matters ranged over all aspects of defense technology, including communications, artillery, antifouling paint for ship bottoms, and many more. And we would talk about ways we might help. We included the information from our discussions in draft packages for an agreement in Washington with the Indian military attaché. The substance was essentially this: Are there technical things with which the US military can and is willing to help? We were, and are still, generally careful about helping people who are not outright allies, and it was obvious that India could not be an ally. But a number of people, myself included, felt that India was the largest real democracy in the world; more people voted there in real elections than anywhere else in the world. If we could not be formal allies, we should at least try to be friends. And that was the spirit in which we carried out our conversations and work.

In addition to the labs, we also went through many military camps where they were trying out new equipment. They wanted to know: Can you help us with tanks? Can you help us with artillery? Can you help us with small arms? Our job was to find ways to help the Indian military with R&D. We would pay for part of it, or we would get something in return – in exchange information or know-how from the Indians, and they would get technology from us. My role was to be an explorer/emissary for the whole DoD.

I am a technologist who likes to apply technology, so this challenge was more like being an entrepreneur – could we get people to use the technology? The Indians wanted help and we were willing to give it in select circumstances, and the biggest "get" for us was the high-altitude medicine and other operational experience. The Indians benefited, too. They made a lot of high-level personal contacts that continued even

Figure 4.38. With Alaistair Buchan, a founder of the International Institute of Strategic Studies, London, October, 1967. CH personal collection.

after the program died (more on that in a moment). They learned a lot – mainly they realized they had missed out on being part of the intellectual defense-related dialogue in the developed world.

I acted on this insight later; I had been a member of the International Institute for Strategic Studies in London since 1968. After these visits to India, I urged some of my British friends who were running the Institute that they should talk to the Indians more. Now, at every meeting of the IISS there has been an Indian sub-cabinet or a cabinet member to give talks, and quite a few attendees from India. That was professionally satisfying, as were my trips to India. But those trips were also satisfying on a personal level. They were very strenuous; I was working all the time with little sleep. But they were intellectually and professionally stimulating, and marvelously exciting.

It took many trips around India to realize how physically vast the country is, how diverse in human ways, and also, in many areas, how impoverished. One odd skill I picked up was the ability to determine the quality of life in a village from the fatness or thinness of the buffalo. I was fortunate enough to always eat well, and so I came to love Indian food. I remember one official luncheon – I think it was in Bangalore, in a huge circular tent, with a long, round table all around the inside of the tent – and I realized that the lunch was all vegetarian. This was to be a new experience – and was even more unique when I sat down and realized there were no utensils or chopsticks! But they showed me how to eat with bread and some spoons and I found the meal absolutely excellent.

I wandered around markets, and spent a lot of time in temples and mosques. In those days you could just go in, take your shoes off, and watch people come in to pray. And you never had to worry about losing your shoes or having them stolen! I vividly remember visiting the Taj Mahal. It is an astonishing place, and the visit a deeply moving experience. After my visit to the Himalayas for a few days, which also was an extraordinarily moving experience, I asked Bhagavatam, "This has affected me a lot. Does it affect Indians also?" He replied, "Yes, it changes your life."

In my experience, people were charming, and only on rare occasions did I encounter any anti-American sentiment. Or it may be that the MOD was more pro-America than other parts of the government. I have always found the Indian military very professional, very friendly to America and to me personally.

From a defense perspective, it was clear the Indians were smart, well-educated people, but with limited experience. The government was serious about defense, and also about democracy. The military was professional, but its labs were mostly behind ours. Yet they were working hard to catch up, and were mastering their unique local conditions. There was a clear overlay of British education, which made a huge difference because everyone could explain things in English to anyone who is literate.

I was especially struck by how India was a strategic opportunity for the US. Here was a country that was different than China – better educated, more industrially and economically advanced, democratic in both institutions and ideology, and potentially our friend. Today, of course, things are very different. India and China have elites that drive the edge of modernization, and both have a vast population that is lagging, but trying to catch up. In spite of ups and downs, our relations with India have been usually cordial. So you cannot think about China as a geostrategic problem for America without thinking about India also.

On my last trip to India, Ranganathan, who was an exceptional host and guide, asked me if there was anything else he could show me. I said, "Well, actually, there is. In all my visits, I have never gotten the chance to see some classical Indian dancing; I like the art form a lot. I am intrigued by it. I want to see more." He said, "Well, there is nothing easier. It happens my younger daughter was last year's All-India amateur contest winner in classical Indian dancing. Come to my apartment; she and her older sister and her husband will come dance for you."

And so they did. They rolled up the carpet, music came from a tape player, and they began. Mrs. Ranganathan sat behind me, telling me what was going on, because it is highly programmatic music. For example, Shiva dances and stomps his foot and that represents a big earthquake, and the music reflects that. It was an absolutely wonderful – a truly unforgettable experience.

The End of the India Program

Our efforts in India culminated with our negotiating an R&D agreement between the Indian Ministry of Defense and the US Department of Defense, with ARPA as the US vehicle. We laid out a program of joint research and we signed a government-to-government agreement, which I signed for the US Government and Bhagavantam signed for the Government of India. We signed the agreement in the kitchen of my house in Silver Spring, during a "diplomatic" party that we were giving, celebrating the occasion. For a while, nothing happened – the agreement made its way slowly through two bureaucracies trying to work together.

But then something went wrong between the India and Pakistan. President Johnson, as I recall, said, "No, we cannot do anything with the Indians or the Pakistanis." And just like that, our program died, shot down in flames by politics. The technical issues were simple, but the political issues controlled events, as they so often do.

Still, I have no regrets about our efforts; the entire experience made an enormous impact on me. I saw how a very professional military does things in a different way than we do, and had the chance to get to know some extremely impressive Indian officials. It was all tremendously interesting to me, and gave me a valuable geographical and geopolitical education, not to mention a wonderful cultural experience.

ARPA Afterthoughts

I have found that many things in life become clearer when seen from a later perspective. This is true about my life with ARPA and later with DARPA. Looking back, a serious question arises about balancing objectives and effort for the short run versus the long run, between tactics and strategy, and, ultimately by considering "What Worked, and What Did Not?" In a way, this book is an attempt to assess for my life, "what worked and what didn't."

Our mission in each ARPA program was mainly technical, with both tactical and strategic consequences. Looking back, I'm most interested in evaluating which things that we did or thought about made a big change in the games we were involved in.

Take counterinsurgency, for example: How did we affect that? We did many different things in Counterinsurgency: There were technical things, some worked, some did not. The M16 program clearly worked, but an attempt to build a walking machine with a man in the cockpit (a program of 1961 vintage), to overcome jungle obstacles, did not. I cancelled that project early on, and (inappropriately) it has disappeared from the record. And the M16 story was complicated, involving detours and controversy.

Some technical things did not work right away, e.g., the laser guided bombs failed the first time round. But they worked later and showed a way to transform warfare by initiating precision warfare. Overall, it was a clear success. And while ARPA and everyone else failed in Vietnam, the ARPA program surely helped the Thai win against their enemies. Most importantly, we initiated a serious study of counter insurgency war, and in a series of Counterinsurgency R & D Symposia, (called "CIRADS") that collected the insights from the ARPA program, and of some military campaigns. These lessons were largely forgotten by the military system but rediscovered during the Iraq war ca. 2006, a delayed success, but certainly no failure in the long run.

In some ways the most successful technology program was the Nuclear Test Detection program. Not only did most of the technical things work, but also the policy and politics came together, and that changed the world strategic landscape. It is hard to be more successful than that.

Ballistic Missile Defense success took longer to reveal itself. Technically successful in many ways, it failed first at the policy level. I believe that we discovered early, in 1966 or so, that a thick defense defending the US against a Soviet nuclear attack was not feasible. But a thin defense against smaller attacks might be possible. (I published a paper explaining the utility of a "thin" defense in *Nature* at that time; 1971.) Later, the ARPA technology played a major role in the Strategic Defense Initiative (SDI), and collapsed at the politics level with it. But "thin" ABM has become a reality now, with the beginnings of a US world-wide "thin" system (aimed at blunting small attacks) with many of the missiles deployed on ships (another ARPA discovery: Sea-based Anti-Ballistic Missile Intercept System (SABMIS) and follow-ons.)

The M16

And it applies, on a smaller scale, to DARPA. A good example is the story of the M16 rifle. We, in ARPA, saw that program as a tactical move to help the South Vietnamese. However, McNamara decided to make the M16 the standard infantry weapon for the US Army. By God, that was a strategic move which changed quite a lot of big things. And now, over 40 years later, we still have the M16 and the many derivatives of the basic M16 design. Our troops are using it, and they are finding it quite satisfactory.

I was happy that the M16 was a tactical success in South Vietnam, and I was looking forward to spreading it far and wide in the tropical counterinsurgency environment. Getting the Army into the M16's story was a detour for me, which I did not enjoy. Sure, you always like your baby to be big, beautiful, smart, and effective. But that was not what we were after – and I got a lot of grief from Army brass for pushing the M16 on the Army, which I never did. I facilitated it because my boss wanted it, but never proselytized to the Army. I was not smart enough to decide that, and nobody asked me to make this grand decision anyway. Strategic moves of that scope have a tremendous political component, which I believe is for the presidential level. That is not for directors of DARPA to try to steer.

Anti-Ballistic Missile Program

But whether decisions are strategic or tactical, you first have to decide whether to focus on the technical outcome or the political, military, and general outcome. ARPA developed lots of ideas and lots of technology about missile defense. I would say the technical success rate was 50 to 75 percent, which is very high for a high risk work like this. But then, we spent a lot of money on it, and we could re-do things that we thought were important, such as Interceptor design or radar design.

But politically, what happened? The Nike Zeus, which the Army was supposed to deploy, was never deployed. The Nike Safeguard (again Army) was never deployed fully. They had everything working at one site for a few months and then shut it down again for many reasons, including a (political) presidential decision to do so. What we are deploying now is a good system for what is wanted – which is to intercept the occasional attack of a few missiles that might be launched by a rogue country, or by an accident. We figured out very early, within the ARPA family, studying various possible ballistic missile defenses, that it would be impossible to build a reliable "thick defense" that would protect the US against a large, well-executed attack without suffering enormous civilian casualties. I think it was the right answer, technically and strategically. It was simply too hard.

But we also realized we could probably neutralize small missile attacks under special circumstances – with defenses on ship platforms, for example. (The whole technique of intercepting ballistic missiles from ship platforms is something that we, in ARPA, invented during the

Cuban missile crisis. Then the task was that if maybe a dozen Soviet missiles were launched from Cuba, one could shoot them down from ships while the missiles were in the launch phase. We thought about it, did studies and many field experiments, and proved it can be done.)

And now, many years later, several of our cruisers have the ability to shoot down missiles and are part of the National Missile Defense System. Is it foolproof? Absolutely not! Would you desperately wish you had the capability, if you came to that? Absolutely. The Israelis have a pretty good ballistic missile defense against short-range missiles, and they are working to extend it. During the first Gu f war in 1991, they shot down a number of Iraqi missiles. We shot down a number of missiles with the Army tactical missile defense system.

There were big arguments then about how effective these intercepts were. And the technical arguments and the technical data have been so politicized, that I think we will never know the real answer. However, the political effect of the Israeli and the US missile defense system in the Middle East war was a great success, because Israel was not forced politically, by the Israeli people, to enter the war against Iraq, which was a huge success – vastly underappreciated at the time. If Israel had entered the war against Iraq, many Arab countries with troops on our side – such as Egypt and Syria – likely would have quit the alliance.

And DARPA technology provided the main technical menu for SDI, which likely contributed to the demise of the Soviet political objectives.

So did the ARPA missile defense program work? Some say "no," some say "yes." I think the "yes" is pretty firm on the technical side, and on the political side also. And because we are now building our national defense system, it is less likely that a threat by, for example North Korea, will change American policies all that much. We will not have to go to war because of the threat. So the DARPA ABM program was a very large strategic hedge, which in the end enabled the US to manage the Cold War Arms race, and left a large and useful legacy for the Post-Cold War period.

ARPA Nuclear Test Detection

Another ARPA program that I also consider to be a major success, technically, and by its contribution to politics was related to arms control. Our aim was to determine whether what looks like an earthquake was really a nuclear explosion underground. Detecting nuclear explosions in space and in the atmosphere was relatively easy, and ARPA had a built a satellite system to do that. Nuclear explosions in the ocean were easy to detect – they cause the ocean to ring acoustically, all over the world for hours – a unique signature. So detecting underground nuclear explosions was the question of the day.

The ARPA NTD program put tremendous effort into research and then deployment of systems to detect nuclear explosions underground. Eventually we succeeded in building a system that seemed reliable enough to detect and identify virtually all underground explosions.

The underground detection program turned out to be a very successful scientific endeavor. And on the way to the policy goal we managed to change geophysics and what we know about the earth by the techniques we developed to distinguish explosions from earthquakes. Bob Frosch (my deputy at the time) and I talked for hours about that. And we said, "Is it not wonderful that in solving this major national security problem, we are doing something major to advance a science we love?" And indeed, the science that became possible with high resolution looks at the earth's interior has revolutionized our understanding of the earth, and other astronomical objects.

From the strategic point of view, the program made it politically feasible for the United States to sign the Partial Test Ban Treaty because the risk was deemed small enough, though not zero. Risk is never zero in the real world. The Comprehensive Test Ban Treaty, which includes underground explosions, has been signed by the US but not ratified. But we have had a moratorium for underground testing for two decades, and this has worked so far.

The Classified Journal

A mixed result followed the establishment of a classified journal. In the early 1960s, it was evident that the written technical reports in the classified areas were of very uneven quality, so there was no reliable way to see whether the work was good or not. We got briefed and we had experts check, so we usually had a pretty good idea, but there was not a body of work that one could point to definitively.

I pushed for a fundamental change, a basic solution for this problem: for a central repository, like the great science journals of the world, that had published refereed papers about work done. What was needed was something similar for the classified world. So, in 1963, I proposed that ARPA start the *Journal of Ballistic Missile Defense Research*. It involved an enormous amount of work to overcome bureaucratic resistance, after all, this had never been done before. The journal was set up. We received papers, refereed them and published them, all under the standard rules for handling classified material. After a year or two, it became clear that people wanted to publish also stuff that was not related to ABM, a lot of stuff related to the war in Vietnam, for example. And so we morphed the journal into the *Journal of Defense Research*, which was a success. It was peer-reviewed, professionally edited and produced, and it was classified, "Secret."

It bubbled along for many years, but died when I was Director of Defense Research & Engineering. Vic Riess, who was director of ARPA then, did not want to support it, and they found that a readership survey revealed it was not read widely. (They did not do a comparative study of how much people are reading the shelf copies of the *Journal Physical Review* – not very many because the distribution is different, because people get reprints or preprints.) The journal died. So now we are back to contractor's reports, which are often just briefing charts that are too busy to use or too brief to explain much, or are mere sales brochures for the contractors. And now people are again saying, "We really need something, because we do not know what has been done." Well, right, surprise, surprise. In my view we now have very little idea of what work has been done in classified technology, because there is no high quality archive that defines the state of the art. For a country that spends billions of dollars a year to generate more classified technology, this is a bizarre state of affairs.

Few now remember what we did and how technology programs were used by the country to manage major national security problems. I hope my effort here and other places will fill in some of the blanks.

Interdisciplinary Materials Research Centers

Finally, one more success story, about a program that had lasting effect on the country, particularly the economy: was the establishment of interdisciplinary materials research centers. The idea came from President Eisenhower (via James Killian). Killian, as president of MIT, supervised one of the two or three greatest materials science centers in the world. He knew that field from the inside, in terms of funding, faculty, student flow, and effective innovation.

But there were some things missing – a vast area of material science that was not vibrant, almost untouched. Metallurgy in 1950 is but one example.

The great idea was that there should be a better, more integrated and more strategic outlook in materials science. When President Eisenhower set up ARPA, a key program that he included was the establishment of interdisciplinary materials centers in several universities. Some stringent conditions were drawn up for universities that wished to establish these interdisciplinary centers. They were encouraged to think about starting these centers with integrated faculties from several different departments, to foster joint research, and devised new integrated courses taught by a variety of people, so all the professors would learn new fields. Students were offered assistantships and fellowships to work in these centers, allowing those who were not wealthy to begin to think of a graduate career in materials science – a degree that did not exist before.

A package was put together describing how to compete to be a member and how to run the centers. The program included ample new funding. It was critical that the contracts were competed, and the competition was fierce, but fair. Fourteen awards were made, and fourteen centers created over a period of about five years. The format of the contract was flexible, depending on the type and quality of the proposal. Some schools got buildings, some did not. Some got a lot of equipment and faculty positions, some got less. The contracts varied in size from half a million dollars a year to 5 or 10 million dollars, which would be 50 or 100 million dollars today – real money. Some established universities were awarded contracts – Harvard, Berkeley, Stanford, the University of Chicago. Others went to ambitious universities, like Carnegie Mellon, which helped bring information processing and computing to materials science.

ARPA managed twelve of the centers, and the Atomic Energy Commission managed two. They started quickly, usually within a year or two. It was a huge shot in the arm of the scientific community, because academics were able to start from a blank sheet of paper and decide what to do. If the history of the program had been properly documented, we would now know how many PhD students were produced, how many papers got published, how many new products were made possible, even how many Nobel prizes were earned because of all these. Estimates run as high as ten thousand PhDs over a period of 30 years. But what has become absolutely clear to me in retrospect is that the electronics revolution of the 1960s, 1970s and 1980s – whose impact is as important as the printing press – could not have happened without this program.

Finally, DARPA as a Strategic Hedge

DARPA is itself a huge hedging strategy against major changes in national security problems, indeed a hedge against the next war, that will surely be different than the last one. It is a rational strategy because DARPA is the best instrument developed so far for understanding what the next war might be like. It will not tell you directly what it will be like, but it gives you lots of novel, relevant, and interesting ideas as to what it might be like. And you have some control over implementing some of those.

But now we have more technology on the shelf than we know what to do with or that we can afford to use. What is missing is better thinking about how major technical goals, like the old projects, could guide broad technology efforts, forcing trends that might change war, and how we can exploit this for our national security. Of course, we should have equally imaginative thinking about non-technical changes in warfare. Who predicted suicide bombing as a strategy? No one, even though we had the Japanese Kamikaze attacks of World War II as a model!

The dividing line between developing technology in response to actual events, as opposed to in anticipation of potential events, is never sharp. Nowadays, it is hard to think of something that should be done that has not been. That was not so in my years at ARPA – when IPTO started, there were no computer science departments filled with smart people inventing cutting-edge technology. Today, the field of active inventors is so much larger – it's a safe bet that anytime we write down a new idea, one of the thousands of graduate students at a hundred of the best research universities is already working on it. You cannot restrict innovation; human societies want to be open societies, no matter what. The challenge is to increase innovation but in way that increases the likelihood of the innovations making a big difference. In that sense, technology strategy is closely related to investment strategy.

A lot of people who worry about these strategic and technical issues have no real understanding of this. I think the only way you can really understand it is if you got your hands and mind involved. Until you have written and published scientific papers, you don't know really how the process of doing science works. And until you have built real things starting with scientific ideas, you don't understand how that process works. There has been criticism over the years that ARPA generated too many papers or did too much thinking. Again, that comes from people who just want to have fun building gadgets, but don't think enough about what they are for.

In my view, people nowadays are not generating enough papers. There is no longer a discipline that enforces reports and papers being written. Now, managers are satisfied with "white papers," which amounts to: "You give me your idea for free, then I will think about whether I will give you money" Or else the results are only "reported" in briefing charts that are not reviewed or archived, and are really not

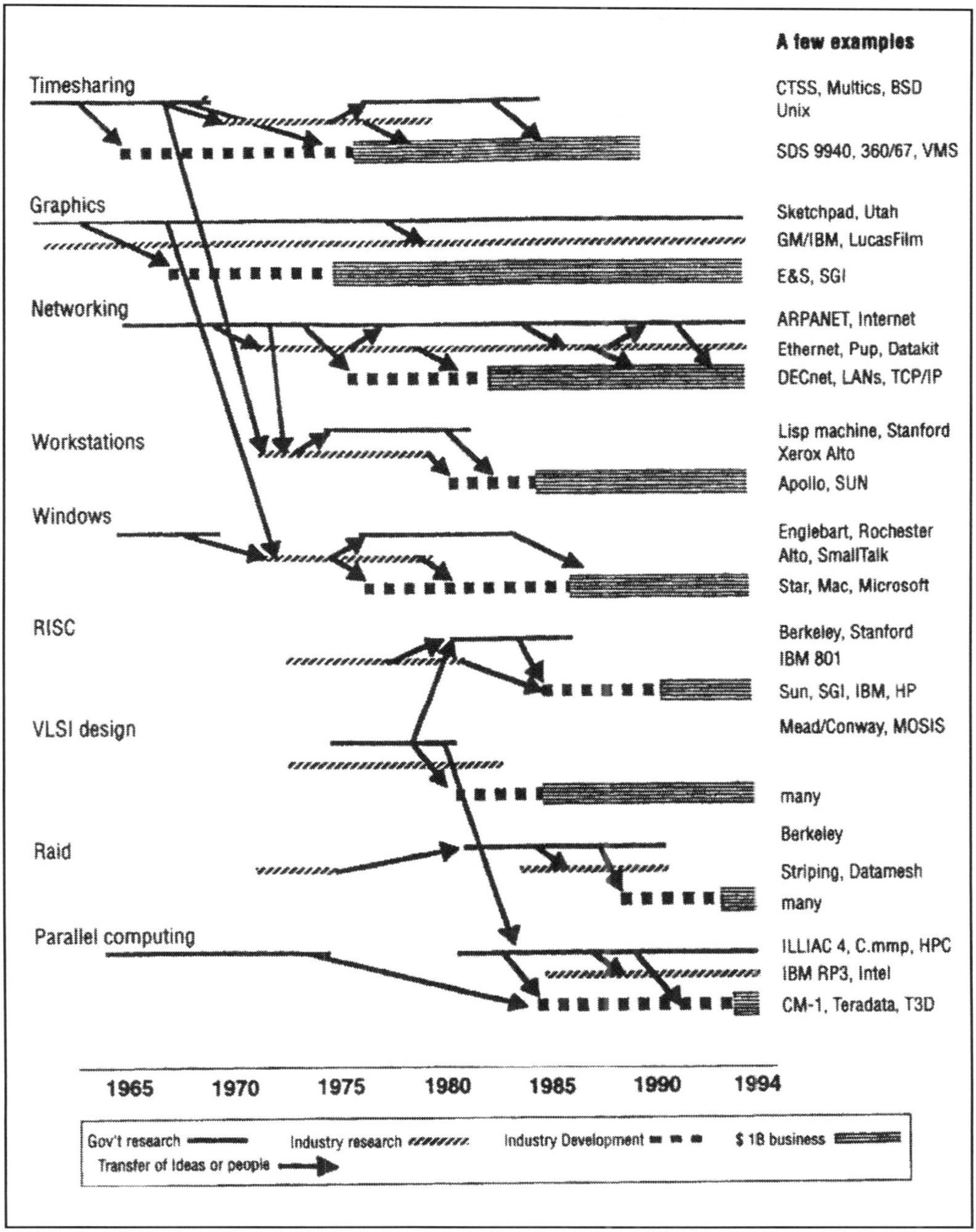

Figure 4.39. Flowchart of work from IPTO in 1965 until near current time. Image Source: "Evolving the High Performance Computing and Communications Initiative to Support the Nation's Information Infrastructure." Committee to Study High Performance Computing and Communications: Status of a Major Initiative, National Research Council. Figure 1.2 p. 20. Reprinted here with permission.

memorable. This problem is particularly severe in the classified literature. One consequence of this state of affairs is that no-one knows reliably what development has been tried and what the results were, and the making of strategy becomes tricky.

But the gem of "what worked in ARPA" was the ARPANET in IPTO. It was a mixed program, part science, part engineering, part human interactions, that changed the world. It is instructive to see how it began to influence all life. This is what a home run looks like.

CHAPTER FIVE

Business Life and the Cold War

Leaving ARPA

In the spring of 1967 I was going full speed at ARPA: the Arpanet program was started, I was going to India to arrange a joint US/India program on military R&D, Vietnam was a challenge, so was the new Thai program, and Advanced Sensors people were becoming engaged in the war. I was very busy; I was also going broke. The ARPA salary was at the top level of civil service salaries, but my duties required expenses not incurred at lower levels of service, but these expenses could not be reimbursed by the government: I had to do some entertaining of colleagues, particularly from abroad, and had to give some major receptions, especially with the Indian program, such as the great signing celebration of the government to government agreement with the Indians. That signing took place at a reception in my house in Silver Spring. Most of these expenses were not reimbursable under government regulations. So, in spring of 1967, I had a long discussion with my boss, Johnny Foster, explaining my dilemma. I had three very bright kids, and their education would be a major financial challenge. He tried to persuade me to continue as we were, but could not really help.

Stories were told that I had difficulties with some of my superiors. Of course, there were disagreements and frictions. But the most significant problems came about because of a perfectly straight disagreement that was partly judgment about how OSD should be run, partly office politics. It began with the request by the Office of the Secretary of Defense (OSD) bureaucracy dealing with office space assignment in the Pentagon, to move part of the ARPA staff out of the Pentagon. I opposed this on principle, because I saw this move as the beginning of making the OSD staff less technically competent, to "outsource" S&T from the Office of the Secretary of Defense. That struck me as very foolish, just as we were developing hi-tech "precision warfare." And contrary to the assurances I received that the move was only temporary, over time it became permanent. All technical expertise was outsourced, organizationally, but importantly, physically the "techies" were moved out of the Pentagon. The consequences of this were just as bad as I feared: In the old way, if a policy person needed an update on a technical issue, say the performance of an IR satellite or sensitivity of a nuclear test measurement, the policy person would call and say: hey Ralph, drop by for cup of coffee, we need to talk. Now it has become scheduling a conference, set up transportation and clearances. R&D has been outsourced. I believe this is destroying the inestimable advantage we had of a smooth interaction between policy "customers" and R&D providers camping under the same roof and only separated by minutes walking indoors.

Anyway, I had to leave for personal reasons and I told Foster this early in the year 1967. Neither Foster nor McNamara wanted to lose me, but they could not fix my problem.

During this time I had a very interesting offer from academia. The people who were putting together the University of California campus at Irvine approached me to see if I would be interested to become Dean of Engineering at the School of Engineering there. Several long meetings with their recruiting people followed, going into some details of university policy, looking at the plans for the faculty and buildings of the Engineering School (fabulous by standards of that time!)That would be a dream job for me. It paid well and it would enable me to have outside income; it was a very elegant offer. But that was at the beginning of

67 and I was really worried about the academic atmosphere, because I was "tarred" by the Vietnam War. At many universities opposition from those who were against the war was a major source of unrest and demonstrations, that occasionally became violent. I did not want to go to a university campus, where I would spend all my time being demonstrated against. So I decided, reluctantly, that academia was out, and I began looking for an industrial job. I was interested in large multinational corporations, because I thought, correctly as it turned out, that the future of the world economy was international in scope. I was contacted by an ITT recruiter who asked whether I was interested in them. Several other companies had made soundings, but nothing looked very serious until the approach by ITT. They invited me to visit the campus of the defense-space group in Nutley, New Jersey, to talk it over. I visited there several times and also went to New York ITT headquarters to meet key people. I remember my key visit to "World Headquarters" in New York. I was going to Europe on ARPA business that day and they really wanted to see me and I really wanted to see them. So I went to New York, and then on to Europe. (I flew from the roof of the Pan Am building , near ITT headquarters, on a helicopter to Kennedy airport to catch the Pan Am flight to Europe. (That was exciting, it used to be a great way to travel.) We had a very successful meeting; in the end they made a very nice offer, to become Technical Director of the Defense Space Group, headquartered in Nutley, NJ. They took my wife and me to New Jersey several times to look at houses. Both my wife and I were appalled at first about the location, "My God, New Jersey," we said. All we knew about New Jersey was the Jersey turnpike which, the closer you get to New York, the worse the scenery is, really not a very appetizing place. But they showed us why they call New Jersey the Garden State. There are lots of fabulous, wooded, country areas, lots of big houses, a very nice way to live. And I was going to make enough money to have a real house in one of those semi-rural places. So we looked at a couple of these and said, "Yeah, we can handle that."

The next events were good-byes. Soon after announcing my leaving DoD, I was invited to the annual White House reception for senior defense officials. Very many people were there, and my wife and I met

Figure 5.1. With General Goodpaster, President of National War College, after a talk on Defense and Space technology, US Defense University. DARPA photo.

Figure 5.2. Fin Larsen, Deputy DDR&E, pins on a medal. DARPA photo.

Figure 5.3. Fin Larsen and Peter Franken, Deputy Director of ARPA and I. DARPA photo.

Figure 5.4. My family, happy that this phase was over. DARPA photo.

Figure 5.5. The early ARPA directors, Betts, Ruina, Sproull, CH, Rechtin, Lukasik, clearly the future was in good hands. DARPA photo.

President and Mrs. Johnson, who were very gracious in what must have been an ordeal for them, so many people, few of whom they knew. But McNamara saw me across the large room, and he pushed his way over to us. “Charley, what is this nonsense about your leaving, what is going on?” he said. “Well, boss” I said, “it is mostly battle fatigue, I guess.” He looked exhausted and tense himself, and he said “Yeah, I know what you mean.” I continued meetings and gave more talks.

After this was an official good-bye from OSD, with a medal for me, and brief comments from me, thanking everyone there, but especially the ARPA people. And for the first time it sank in on me that this was a huge change in my life, and also how much I would miss the ARPA crew, and how I had not expressed adequately how much I cared for them, their efforts, and their future.

The ITT defense-space group was dealing with many products, such as ship radars, avionics, navigation aids for civil aviation, Army communications; things I had had nothing to do with at ARPA. So all thought there would be no real conflict of interest problems, I only had to stay away from government committees for a year. They moved very quickly, and by end of May I was working at ITT. At ITT we had constant meetings (and lots of committees), but the big meetings really decided things and decided things quickly. Having large meetings was ITT's way to get transparency for the management, it took a lot of time to get the facts, but once they got the facts, they moved accurately and very fast.

Starting at ITT

Working at ITT was a great education, I learned a lot about modern business. My area of responsibility was to ensure the technical quality of our products and the future development of new products. That turned out to be the supervising of the internal R&D programs, as well as of the advanced development contracts. They sent me to training courses run by the company. An example, a very high-level course in accounting for executives provided one of the most interesting and useful things I learned.

I also took courses in human behavior; then called "sensitivity training," except ITT did it differently than some other organizations. They had their own people do it and there was some challenging stuff. I attended several, probably four or five of these, over the many years in ITT. It was a revelation to me, they documented everyone's behavior in our groups, and I got to see sides of me that I did not like; I was too overbearing, too pushy and so on. It changed my behavior some, I think, and for the better. But it was hard to take.

They had courses in how to manage; they had courses in marketing, how do you think about markets. Most people confuse marketing and selling; selling is really a different function. Marketing has to do with business strategy and how do you tell whether a particular field is a good business to be in or not. It is related to financial strategy, not to the accounting part, which is lower level, but is more strategy;

what kind of customers do you want to have, and whether you have a real choice.

The main thing was that I really knew about technology and I really knew about defense problems. I was in the middle of defense and defense policy discussions in my DARPA years and at that time ITT was somewhat remote from those things. Some people knew this, wanted to change that and were happy to have me there; but some people liked the status quo.

I brought broader insight into what was really going on in the defense world, and some people were interested in using that insight and some were not. Harold Geneen, who was then chairman and president, always liked to talk to me; he always tried to learn something new. He said to me often, "What are you doing? Tell me about something new!" Once in the middle of a general management meeting with 250 people around a large table in a very large room (and I sat nearly across from him), he said "Charlie, tell me. Why are we (the US) doing the space shuttle? What is going on here?" And I said, well, basically it is to make going into space and working in a space craft more like flying Pan Am than flying in a metal can, which was what the early space capsules were. He said, "Oh. That is interesting. Tell me more." And I did, later.

The Defense Space Group (DSG) had about $400 million sales a year in 1967 dollars; that is maybe an equivalent of value of three billion – not a small organization.

My duty was to learn about what we were doing in the defense-space group and analyze the technical programs, R&D programs, in the group. My boss, the group executive, was a man called Jack Vollbrecht; a brilliant man. He came aboard about a month after I did, and he very soon asked me, "Charlie, what do you think of the group? Tell me the big story." After some reflection I said, "Briefly, the labs are too orderly and clean and the factory is too messy. So I think we will not be very inventive and we will not be able to make products anyone will want to buy." He said, "That is pretty serious. Show me." So I did. And rapidly these things began to improve under his management.

He took a few of us in hand for "tutoring"; one of them was Rand Araskog, another one was me. And he would sit down with us every

Figure 5.6. My office in Nutley, NJ. CH personal collection.

week for a few hours and he would explain his ideas about managing to us and we would debate the issues. He had written out notes how management is supposed to run; what are the big principles – very clear, fabulous stuff. And I still have the notes, they are still relevant. (Araskog became Chairman and CEO 12 years later.)

At the same time I started to do a lot of technical reviews of the R&D programs. I went to the seven divisions in the group every month and reviewed all their technical programs really in detail. I had a staff of about five or six people; some of them were very, very good and they helped a lot. They were experts in the regular ITT defense products, while I was not yet, it was a great combination. Also, I participated in the management meetings and many marketing strategy meetings and helped to assess whether the plans were sensible. Technical and marketing were beginning to work together, as they must, or much effort would have been wasted.

Figure 5.7. Review at Gilfillan, the ITT radar company in Van Nuys, CA, First row on left, Rand Araskog, Group Executive (later Chairman and CEO of ITT), Bob Olson, President of Gilfillan, second from left. Photo courtesy ITT.

Sometimes I would have to say, "I do not think we know how to do that." It did not make me popular. Or, "I do not think you really see what that program is about. I have not heard any explanation of it that makes any sense to me. I think we had better dig some more." Vollbrecht thought, and I very much agreed with it that if you really are a strong organization, you can turn down some jobs because you know the job is not good enough for you or you are not good enough for the job.

When I first came to ITT, they got low grades by the government people rating our R&D programs, when I left the group after some time and went to do other things in ITT, we were in the 80 to 90 percent approval range, a very significant improvement that was widely recognized, and had immediate favorable financial impact on the group.

Figure 5.8. My family in New Jersey, 1972. CH personal collection.

Figure 5.9. "Cookie" my secretary for many years, celebrating my birthday, and her then husband. CH personal collection.

ITT was very active in night vision devices; the defense-space group was one of the big manufacturers of early night vision devices and they still are very active in this product, are still among the main manufacturers of that.

Well, we put the Group R&D program in good shape. We also started on fiber optics in the 70s and we had a plant that made fiber optic cables in Roanoke, and for a while we were the main supplier of the military for fiber optic cables.

ITT businesses ranged across a wide menu and across a lot of different countries. In its day it was one of the great conglomerates, and I found learning about that an extraordinary experience. The core was telecommunications. By that time ITT was out of the business of running telecom service companies, but it was an important equipment supplier worldwide and had big export businesses, it had major presence in Germany and France and Britain and Italy and Spain. These all sold equipment all over the world, to Europe, Asia, Africa, Latin America, and we had a pretty healthy, and a growing telecommunications group in North America.

The second biggest part of ITT was the Hartford Insurance Company. In fact, later I found out that they employed more computer programmers at Hartford than in our telecom business, which was an amazing fact. We also had products like Wonder® Bread, Sheraton® Hotels, and Avis Rent a Car®.

The defense group management was so successful that the group grew by being assigned other ITT companies, such as electronic connectors, several energy companies, and finally ITT Telecom North America. The group had a number of chiefs. Vollbrecht was one, right after him came Rand Araskog who was the group vice president – and I became technical director of each of these as the group grew, which gave me new fields to learn, which I enjoyed. And the energy work was fascinating; though it never got as large as we thought it would.

And I continued learning new science and engineering, and new things about business. I traveled a lot and got to know parts of Europe very well. I also had fun. I have fun when I can do something about solving problems, and that has been the case all my life. And, boy, are there big problems in a good company. They range all the way from what is in the head of the top guy, to if the lawyers understand what you are really doing. Do the technical people know what is really needed and when and how? Do the engineering folks really build what is needed by the customer? There are endless problems of understanding, making sure people understand what is needed.

That means convincing people. That means explaining and arguing. That means endless meetings. That means you had better have the best facts, accurate facts, and better ways to explain them. But the way I like to work, I actually try to do all that. You do not just write memos; you actually call the meeting and say, "Look, we've really got to solve this problem. You guys and I have got to decide about this."

Importantly, I was making enough money. All my kids could go to the best schools they could get into. I had a very nice lifestyle. I had picked up scuba diving and wound up doing a lot of it.

The Cold War Continues

My involvement with the Cold War went in parallel with the work at ITT and it absorbed a major effort on my part. The first time I ran into the Cold War in a personal way, is described in Chapter 1, when I was coming out of Europe in the middle of World War II. During World War II it became clear that the Soviet troops, wherever they went, had no truck at all with the Allied governments in exile, but set up their own, Soviet-run governments. That happened first in Poland after the dreadful Battle of Warsaw in 1944. That year the Polish Home Army rose against the Germans, while the Soviet army was just across the river. But the Soviet Army did not help the Poles a bit, and refused to let Allied planes drop supplies to the Poles fighting in Warsaw. And the Germans butchered the Polish Home Army. And then the Soviets installed a puppet government in Poland and took over.

After the war ended, the new, anti-Nazi Hungarian government was set up and the Romanian, the Bulgarian and so on. They were all sort of two-step affairs, with governments that were set up initially in the liberated countries, that were broadly based, and at some later stage the communist factions took over the country with help from the Russian army, and suppressed the other parties.

The next personal Cold War event was in 1950 when North Korea attacked South Korea, without any provocation. It changed my life. I went to the Army for that war, as reported in Chapter 3; it was part of my long education. It was clear that the North Koreans attacked first, and after near disaster, the US wiped out the NK army, even though the North had massive Chinese help. And second, they attacked the South only because the Chinese encouraged or permitted them to and Stalin did not object either. If Stalin would have said *nyet*, it would not have happened, because he had really control over these countries.

What really got American attention, however, were the Berlin crises. They were clearly fomented by the Soviets and often by people like Khrushchev personally. These crises started in the '50s and '60s. In fact, the Berlin Wall going up was the triggering event to get me into defense work at ARPA as reported in Chapter 4. And the Vietnam War was a big aspect of the Cold War.

I would say, basically, my whole professional life was caught up in the Cold War. After the Korean War, my contribution to fighting the Cold War came through working in defense industry, ITT, and by participating in several committees. Of these the Defense Science Board (DSB) the Defense Policy Board (DPB) and the Chief of Naval Operations Executive Panel (CEP) were the most significant and interesting.

I first got involved with the DSB when I was still at ARPA, in the mid-60s. The DSB reported directly to the secretary of defense, and existed to give him technical advice. I got invited to meet with them, and got invited to their meetings and summer studies. We talked to them in formal and informal presentations. I got a lot of help from them – learned a lot. It was a test bed for my ideas that was ready at hand, and they were always eager to hear what we were doing and had strong expert views about everything. The chairman at the time I was in ARPA was Fred Seitz, a famous solid state physicist. (He got the Nobel Prize in physics for his scientific work.)

He was also a colleague from before; he and I had interacted with each other on some of the science I did, and he published a long paper of mine, written with Paul Meijer, on how to use group theory in a basic way to determine the wave functions of some systems in solids. The paper was published in a series of books about Materials Science which Seitz edited.

Working the Cold War, "Part Time"

After I left ARPA, I could not join any advisory Defense board for a year, but in summer of '67 – I was invited to join a special meeting of the DSB, about twenty of DSB people and a few outsiders like myself to discuss the strategic balance between the US and the Soviet Union. It became clear that the Soviets were building more ballistic missiles, and bigger ones with more throw weight, an ultimate measure of threat, and the issue was what the US should do about it. After that, in the end of '68, I got invited to become a member of the DSB. The DSB is a really a creature of the secretary of defense and it does basically whatever the secretary wants. If the secretary does not want any tasks done for him, he will not get anything much from the DSB, but others spring up with suggested tasks to keep the gears oiled.

In those days, the secretary of defense and deputy secretary of defense were very actively working with the DSB. They gave us our problems; they discussed the work with us through the year. The board usually met once every month or two, and had parallel panel meetings that were after the specific problems, and they all met for a month in the summer for the summer study. Those were pretty high voltage and high energy operations. And we had to work on very big problems.

Ocean Surveillance

The first significant thing that I remember that I did for the board was a study that I chaired in 1970 and '71 to examine improvements for the US Ocean Surveillance System. That was really an exciting year and the Navy got on board very quickly. My working with Admiral Zumwalt, the Chief of Naval Operations, on his CNO Executive Panel (CEP) helped with this process. We had a similar and parallel activity in the CEP; and all that made a big improvement in the Navy's way of operating and way of thinking, a change and improvement that I think has been permanent. Members of this study included Kent Kresa, formerly of DARPA, later boss of Northrop Grumman; Tom Evans, of the Applied Physics Lab of Johns Hopkins; and many other experts. Also, through the CEP, I found a great admiral, Tom Davis, who was an expert in surveillance technology and tactical use of it. He had recently commanded a significant part of the fleet that was in constant contact with the Soviet Navy, he was an expert at fooling them and outsmarting them, generally. So he was an important member.

At that time the Soviets followed our US Navy ships, airplanes constantly all around the world,. It was a huge effort that involved perhaps a third of the Soviet Navy all of the time, just as a third of our Navy was at sea all of the time.

Gene Fubini, (former Deputy Director of Defense and Engineering) and a very sophisticated friend of mine asked me once, "Why do you work so much with the Navy? You did work with the Air Force and the Army advisory boards. What is special about the Navy?" I thought about it briefly, and I told him, "Look, the Navy are involved with their potential enemy all the time, they track the Soviet Navy, they study them, they understand them. That is their daily bread. It is more interesting, and I hope, more important." And it was more fun.

I had another ocean encounter that summer; I got certified as a scuba diver, and that changed my life in a number of major ways, as related in Chapter 3.

1970 was also a most exciting year for me professionally. Three things happened in that summer. First, I was invited by Henry Kissinger, who was then President Nixon's National Security Advisor, to join a small quiet, nearly invisible group, to help with some major national security issues. Members of this group included Jim Schlesinger, who was then deputy-director of the Bureau of the Budget, in charge of all the national security accounts, later to become secretary of defense, director of CIA and chairman of the AEC. There were other interesting people on that board as well, such as Bill Kauffman from MIT who was an expert in defense economics.

Second, I started the DSB Ocean Surveillance study for the DSB, and third, I was invited by Admiral Zumwalt, newly appointed CNO to join a small group advising him directly, the CNO Executive Panel mentioned above. Each of these efforts involved me in top level work on related issues, but for very different customers. In the outside,

Figure 5.10. Meeting with Admiral Colbert, USNavy, Commander of NATO Forces South, at an Ocean Surveillance Task Force meeting in Naples, Italy. CH personal collection.

Figure 5.11. Tom Evans, an old buddy from Ocean Surveillance. Photo by Susanne Ollmann.

real world, the Vietnam War was still on; it was to go on for five more years. Everybody was trying very hard to not lose it, and we got close, in my view. That's part of the Vietnam story, is told in chapter 4.

We continued in '72 to study the Ocean Surveillance, but this time we analyzed the Soviet Ocean Surveillance System – and how it worked. It was a major net assessment, where you look at US capabilities and at hostile capabilities, trying very hard not to mirror image, not to ascribe to the other guy what we were thinking and doing. We were trying to find out what they think they were doing and try and understand that, and that we did, and it made a good deal of sense.

Soon after the second Ocean Surveillance study, we did a study that I chaired, on "enigmas in Soviet R&D." At that time there was the problem that we understood some of the Russian defense R&D, but not all of it. And there were discussions and speculations about how much did we really not understand. I thought that the stuff we did not understand was much more than most people thought, so I proposed we conduct a really good study of this. At the end we managed to explain a number of enigmas and a sufficient number of that drew attention.

For example, we showed that every time we observed a new ballistic missile silo at the Soviet test ranges, a new ballistic missile would appear on the test range and would be tested later. So the silo building was the leading indicator and they always built a new missile for new style silos. There were other things like that, too numerous to mention here. In the end, it turned out the fraction of Soviet R&D that we did not understand was quite substantial and that had serious policy implications.

A large part of the Soviet R&D was for the huge land army, which the Soviets had in Europe and which absorbed a really large part of their budget and everybody's attention. To counter this threat, the US eventually developed precision warfare, a combination of technology, training, and doctrine that made it likely that most of our bombs and missiles hit their targets. That was a "game changer" that created the so-called "revolution in military affairs." ARPA, the DSB and others made major contributions to this. And I was involved personally via the DSB and other studies.

The second big threat was the intercontinental ballistic missile (ICBM) threat that the Soviets were continuing to build, while we were just finishing our program, we never really had any large, new programs after that. The Soviets just kept building and by that time they had overtaken us in throw weight (which is the amount of weight each one can carry to its target). We had done well with missiles with smaller throw weight because we opted to go for smaller and more accurate warheads. Furthermore, we had the technology to build powerful warheads that were relatively light.

Admiral Zumwalt became chief of Naval Operations in 1970, a move that many applauded except by some old Navy hands because in their eyes, he was deemed much too young. He had run the US Navy in Vietnam, had basically invented the Vietnamization process of getting the South Vietnamese take charge of the war in their defense, which I thought was very much on the right track, and which President Nixon adopted as his major strategy. Zumwalt wanted to make changes in the Navy, and it was clear he had the President's support. The CNO used the CEP to get advice on all kinds of Navy issues, technology, strategy, long range planning, and everything else that came along. It was very private, so everyone involved could be totally frank. And we worked well with the Navy staff.

The Navy and the White House

The CNO found that much of the Navy was very much stuck in the pre-Atomic Age, and he wanted to fix that. The submarine Navy had taken on nuclear energy and carried nuclear weapons, but the rest of the fleet did not want anything to do with that. Albert Wohlstetter, a CEP member and a leader in nuclear strategic thinking worked with this problem intensively, focusing on minimizing collateral damage while assuring strategic utility.

Kissinger asked me to think about what I thought about the future of the US Navy – what would become most important: big ships, small ships, how many, what kind. There was a big political attack on our big aircraft carriers at the time, and many people wanted to get rid of them. But what those people did not understand was that the big carriers were and are one of our crown jewels, the secret weapon that no one else has managed to do as well as we have. They are incredibly effective at what they do, and have been occasionally decisive in geopolitical and geo-military crises.

The three studies, my DSB Ocean Surveillance task, writing a paper on the Navy for Kissinger, and working with the CEP were really not connected, except that they were bound to collide unless care was taken. The ocean surveillance study for Defense Science Board, which was for the secretary of defense, was just pure win-win for everybody, but the Zumwalt and Kissinger efforts had the potential for real conflicts. Kissinger was wise enough never to let us know where he stood on any of these issues. So we just gave everybody the best shots we could, and they would decide what to do with it, which is how it should be. And no disastrous conflicts arose.

Early on Zumwalt asked me to see a group at the Naval War College that was doing a small private project for him, and I did. These were very young officers, commanders typically, who had just come back from the fleet, particularly the Mediterranean, which was most involved with the Soviet interactions. Their project was to construct a draft tactical manual for the use of the Harpoon missile which was just coming into the fleet then. It was our first modern cruise missile for the fleet. It was relatively short-range; it did not have a very big warhead, but it was there.

So these young men showed me, crawling on the marble floor of a big hall at the Naval War College, with large maps and charts and drawings, what they had done. They showed me the draft of tactical instructions for the use of this missile, which was being put on our ships, and it was very interesting. So I reported back to Zumwalt: "This is really good stuff. You are right." And a draft tactical manual went to the fleet early and was usable. Of course it was revised and improved later.

Another thing I did for Zumwalt later, was to look at whether high-powered lasers were ready to go to sea. I was of the opinion in '74 near the end of his tour, yes, they almost were ready and what you should do is get a ship that was going to get sent to the bone yard and devote that as a platform to put a strong high-powered laser on it and take lasers to sea. Because my attitude was and still is if you take anything to sea, it works differently than if you try it out in White Sands, in the desert. That effort of mine failed. Various people killed it, and it has taken the Navy a God-awful long time to get back to it.

He made one major mistake (as far as I can tell) – he redesigned the Navy uniform for the enlisted men, and proposed that everybody get rid of the pea jackets and bell-bottom trousers and Navy hats, and go to a sort of business-style suit with a cap that had a white top and a visor, more or less what the petty officers, the noncommissioned officers, wear. The sailors hated it, and the idea was dropped.

A few times my working with the Navy was strengthened by my scuba diving. I went diving with some Navy guys and the Navy dolphins, an unbelievable experience. To see the dolphins work under water with their handlers, to see them look you over, and go on with their task, was quite wonderful. I asked a Navy enlisted man, who was a handler of the animals, "How smart are they really?" (It is a big scientific issue. They have a huge brain, and we know now that most of that is used for

Figure 5.12. Meeting Werner von Braun, ca 1970. Photo courtesy ITT.

Figure 5.13. CH at Apollo 9 launch. CH personal collection.

Figure 5.14. CH getting into British Aerospace Harrier Jump jet for a flight, this was a rare two-seater configuration. CH personal collection.

Figure 5.15. Meeting with Admiral McCain, the Commander in Chief Pacific, 1971, reporting on various developments. (His son, now Adm. McCain (USN retired), was then a North Vietnamese prisoner of war.) CH personal collection.

extracting information out of the sound that they generate and that they hear. That takes probably four-fifths of their brain structure.) So I asked this guy who really knew them, a very nice guy, very friendly, "How smart are they, really?" He answered "Just about as smart as a smart puppy, very playful, sometimes mischievous. And you can get them mad, in which case, they will bite you and you will not like it, (they have sharp teeth and very strong jaw muscles). Sometimes, they just "walk" away from the job as if to say, 'I'm tired. I'm not doing this,' and go to rest somewhere. They find work very stressful. They like to play. They play endlessly and want to play with humans."

When they do play with humans, one of the problems is that they are very, very big. They have no idea how fragile we are. So accidents happen, they may break your ribcage and not know it and not want it. They are appalled when you get sick from their games.

After a while of work, we were able to advise the Navy on their mammals program about various problems.

I found the Navy as people and as an organization exceptionally dedicated and professional. I have been proud to work with them, and to be able help them. This is also true about the Marines. When I became DDR&E (to be discussed in Chapter 6), I had a choice for a military assistant and I asked for a Marine. People said why a Marine? "Well, two reasons. One, because that job is full of potential professional conflicts with the services. A Marine working for me was less likely to get into a big fight with his service about weapon systems, because they do not buy much of that." "What is the other?" I was asked, and I said, "Marines, if you ask them to do something they will say, 'Yes, sir' and go do it. But they will also say, 'Sir, if you are going to do what you want to do, you are going to crash.'" And I liked both reactions.

Meanwhile life at ITT kept up its normal pace.

Going to New York Headquarters of ITT

In 1979, I came to work at New York headquarters. Rand Araskog had left our Group and went to ITT headquarters, and in late 1979 he asked me to move too. With his guidance and encouragement we began to attack the deeper problems of ITT technology. Two issues dominated: the advanced microelectronics technology then required in some of our products, and a lack of vision in some of our technology programs. Araskog and I began to develop a comprehensive strategy for attacking these problems. However, before that effort got going, a major hiccup happened that turned out to be a long term blessing.

ITT then had twelve large labs, ten in Europe and two in the US. Once a year we technical folks had a whole day with the ITT board of directors to explain what the labs were doing and why; it was "lab directors' day" and it was typically in January or February and it was always held in Brussels. The whole board always came to these meetings.

I worked on the first one soon after I got to New York. I was in charge of the program, and the meeting was a total disaster. The talks of the lab directors were not understandable by lay people such as the board members, and most of the board members went to make phone calls, or held small side meetings – a total disaster. At the end of the day Araskog said, "Charlie, it is never going to happen again, right." I said, "Yes, Rand, it is not going to happen again, ever."

We had lab directors' meetings several times a year. I chaired these and set the agenda. And at the next one I took the floor and said, "Guys, never again, next time we will do it differently." Then we spent several meetings about how to fix it. The interesting thing was that the lab directors, mostly Europeans, were very good technical people; but they had no idea how to talk to a non-technical group or a Corporate Board of Directors. So I told them, "Look guys. You do not have to prove that you are very good technically. You got your job because everybody thinks you are. So you don't have to, and are not allowed to talk about technical details. You have to talk about what are your ideas and why anybody should care. It is not your PhD final; it is something else. And you need to learn to do this kind of presentation very well."

They were kind of annoyed with that, and did not believe that this was the problem, they blamed the audience. To make sure that we had the right approach we had three dry runs before the next meeting with the Board; one was about three months in advance, when we set the general objectives of the meeting, and went over every presentation, tossed out a lot of slides, and so gradually began improving the

Figure 5.16. Charlie Kao, inventor of Fiber Optics, my successor as ITT director of research and technology, Nobel Prize in physics, 2009. Photo courtesy ITT.

talks. One dry-run was one month in advance, and the last one was the week before the meeting with the Board. They all said, "We do not need that last meeting." I said, "Yes, we do, there are still a lot of soft spots, we have to fix them. I will tell you how serious I am: We are going to have our last dry run at the London Heathrow Sheraton, and I'm not taking a room. I am coming in on the red eye plane from New York, then we have our dry run, and I am leaving on the evening plane. But I am going to look at every slide of yours and I will pull out the slides that are too detailed and too technical, and we will mostly leave the "why" slides. You still are trying to sneak in material that is too detailed, and that has nothing to do with the bigger subject: "What are the Labs doing, and why."

Well, the meeting was a great success and the Board members stayed for the whole thing, asked questions, and were happy, and so was Araskog. The Lab directors gave me a dinner afterwards. They carried me into the restaurant, literally lifted me and carried me in. They had never been so well-treated by the corporation before because, they had learned how-to-talk-to-the-board, which I knew and they had just learned.

Key Technology Steering Groups

The systematic attack on our technical strategy began with the formation of what came to be called the Key Technology Steering Groups (KTSG).

The real task was to revitalize the vision of the technical department.

This was a very important professional challenge for me and it had to succeed. So I spent six months crafting the approach. I worked with Araskog and talked to him every few weeks about the various approaches, and, finally it looked right, and he agreed to do it: To set up the key technology steering groups, and run new programs through them. In the end we had about eight of theses groups, each had an assigned area, like micro electronics (which was the first one set up, it was chaired by Barry Soloway, a renowned expert in the field); the second one we set up was for fiber optics (chaired by Charlie Kao, who invented modern fiber optics in 1967, and got the Physics Nobel prize for that in 2009), a third one handled advanced Software, chaired by Tom Love, and so on. Their function was to devise an advanced corporate R&D program in their area, and help build a suitable corporate market strategy approach, then to set up the programs, and finally to actually guide and run the programs. The name for these groups really mattered a lot, it took months to find the right name, and “Key Technology Steering Group” said quite literally what we wanted them to do, so that made it the best name, in spite of the ungainly length of the name. The abbreviation, KTSG, became a trademark of sorts, showing that good ideas can have bad names and still succeed, but it was a risk.

And I managed to persuade the corporation to get extra money for the R&D programs devoted to these areas. We started with on-going relevant programs of about $30 million a year. At its peak, four years later, the advanced technology program was about $120 million a year; those were big increases.

I insisted that the KTSGs be international; by that I meant that the membership was roughly half Americans, half Europeans. I insisted that we meet every month because we had a lot of work to do. We had to devise programs and then run them. I also insisted that our meetings would alternate between the American side and the European side of the Atlantic.

That made a hell of a lot of travel for everybody and a hell of a lot of travel money. The top management was very uneasy about the cost, but they finally went along with it, and it worked. I said, "If we do not do it that way, it will always look parochial and that will fail because the Europeans will say, 'Oh, the Americans are not listening to us'." And I wanted to break the back of that argument.

The Lab directors were key players; they had most of the people and the resources for these key technology steering groups. So the lab directors helped with the management, and they got most of the new programs.

We got the steering groups started in 1980, their first year. And in the end it just worked fine, and whenever anybody wanted to have a detailed discussion about something where there was a steering group, I said, "Talk to the guy who is the chairman. He'll know more than anybody in the company."

The steering groups had a mix of technical and marketing people. We wanted the best person in the company who had the job that was in charge of that problem, the best person technically or in marketing. In the first few steering groups that we set up I personally picked the chairman; he got vetted by a number of people and I said, "You and I are going to pick all the other members and I want to get to know everyone. And if I do not know them now, let's arrange that I get to know them." And that is what we did.

It involved a total of maybe a hundred people; they were the best in the company, and it was pretty exciting. Sometimes they did not know what to do, but we would figure out how do it – build a program and run it, review it, change it, build a next-year program, argue for the money for the next year – like "doing" DARPA stuff. It changed the company. When the Corporation later sold the telecom part of ITT to Alcatel, most of these companies that were involved in Europe were sold to Alcatel. Alcatel kept these steering groups for several years after.

It was fun. I worked like crazy, but I had a ball.

Interestingly, Araskog had been in ARPA early on. He was an early member of ARPA; he was in uniform. He was an Air Force Lieutenant,

He led most of the members of the Software KTSG into East Berlin for dinner one evening. We got out freshly from West Point; the Air Force Academy did not exist yet. He was Roy Johnson's, the first director's, special assistant for space programs. So he coordinated all the early ARPA orders about space. Therefore, he knew the model I was replicating at ITT.

But not all was lab work. At that time we had an interesting job to build satellite communications ground stations for Indonesia. The Indonesians were the first to have a domestic communication satellite, called Palapa, which they had put up in 1980 or so. And in '85, before I left ITT, I went to Indonesia to check out some of the ground stations we were building for that system. I foresaw trouble doing it because nobody in our group knew how to build in the tropics, which is really a different challenge. However, from my ARPA experience building things in the tropics I knew what to look for. So I went and looked for future problems, and I found some. For example, there were a lot of bridges in Indonesia between the harbor where the station electronics would land, and where the station was supposed to go. The bridges were lousy and were likely to fall down. So I went and checked them out and I took a lot of pictures of the bridges.

When I came back, I said, "Look, all you have to do is get a gang of workers before everybody knows what you are doing, so it is a little bit cheaper. Buy about thirty large pieces of good lumber; do that also before anybody catches on so it is cheap. Get a couple of trucks in addition to the trucks that will carry the equipment. And you put that lumber down on one bridge and strengthen the bridge, take your equipment across the bridge, then carry the lumber to the next bridge and build on that bridge and carry it to the next bridge, and so on. You work your way to the site and set up our equipment." Well, the equipment went to right spots.

Cold War Again

And the Cold War kept going. When the '80s came along, the DSB let me know that I would not be re-appointed, I had been a member for a long time. In the prior year, I had been introduced to the Defense Policy Board, which was a totally different organization. It was newly reconstituted and full of high octane people. At first I was the only technical person on it, later others were added. My involvement with the DPB was a wonderful seven or so years, '83 to 1990, but I had to quit the Defense Policy Board when I became DDR&E in 1990.

The DPB had been dormant, but was revived in the Reagan days, and the secretary of defense was Caspar Weinberger. He worked very closely with President Reagan and was very involved in the strategic thinking and strategic buildup. The DPB was to help the secretary with policy issues, and we did.

The members were mostly policy people. The chairman was Seymour Weiss, a senior State Department career official who had served in several ambassadorships, and was a member of the CEP as well; he was a wonderful personal friend. Albert Wohlstetter was a big factor in the DPB. There were several retired military: John Vogt was a retired Air Force four-star; Admiral Bill Crowe who later was chairman of the JCS. A lot of very well-known political science people, Ash Carter, at Harvard, now Deputy Secretary of Defense, political experts, strategic experts. I discovered in the first meetings that the DSB and DPB had not one single agenda item in common – not one. Nothing that one board was pursuing because it was important was reflected at all in what the other board was pursuing and thought was important. (An obvious candidate for the agenda of both boards might have been the question: why do our weapon systems take so long to create and are so expensive? In fact that would be a good challenge question to put to both boards today, but jointly.)

I think that is highly indicative of a major problem of government. It is symptomatic and it shows the split of the "Two Cultures" described in the book by CP Snow. Eventually they got some technical people in the DPB, first John Deutsch, then John Foster, and later Bob Everett, but the crossover did not work the other way around. Technical people are uncomfortable with policy and politics, and policy people are uncomfortable with technical things and people. That is a cultural fact of modern life and it is awful, and also very dangerous, and it bodes ill for our futures.

Anyway, a number of important things were going on in DPB. Weinberger had started the publication of an annual DOD brochure called *Soviet Power*. It was a breakthrough. It was full of facts and pictures that had been highly classified before. And it said some real

Figure 5.17. At the end of the work of the Commission we presented our findings to the President and all met with him. Photo courtesy Ronald Reagan Library.

Figure 5.18 CH with Herman Kahn, a continuing relationship. (I wonder why I was so intense in the picture?) CH personal collection.

things about the Cold War and how we thought we were managing it, and how we thought the Soviets were managing it. The publication was very carefully done, and over the years served a huge US policy purpose and helped avoid World War III in the '80s. The Reagan administration was able to explain through this publication some very important issues in a number of ways. And the Defense Policy Board was working with the Defense Department on the *Soviet Power* annual publication. It was one of the media that began to explain to the leaders of the Soviet satellite countries that if there was a war between the US and Soviet Union, which would presumably be nuclear, if a country like Poland, for example, supported the Soviet Union, Poland would suffer terribly. However, if they tried to slow roll their support of the Soviets, we would try to spare them.

It was only insinuated; it took several years' preparation to get to clear statements of potential changes in policy. But it came out in writing, and that message was also sent in lots of other ways. There was real communication going on.

Another topic covered at every meeting was the war between Iran and Iraq, which Iraq started, and which we were watching carefully. Every meeting started with a two hour briefing about the war. Intelligence people came and told us everything we knew of what was going on in the war. All the data were there and the questions that we asked, we got answers to. It was mighty interesting.

One of the things we discovered was that the Iraqis were following the Soviet "combined arms war" model to the letter. They had Soviet equipment and Soviet training; they had the Soviet organization. They did the Soviet thing and it worked for the Iraqis beautifully for about three months. Then they got stuck and they never got unstuck. It was enormously interesting because we knew a lot about what we thought the Soviets would do (we had taken many countermeasures for the US) and so it was fascinating to watch the experiments being done – both working, not working and to consider why. The war became a stalemate for a number of reasons, limits in Iraqi logistics, fierce resistance by Iran, and failure of the Iraqis to interdict the oil tanker flow from Iran. Finally, the two sides gave up.

Shortly after I was invited to join the DPB, President Reagan appointed me to the new National Space Commission, which was set up by Congress. Its chairman was Tom Paine, former head of NASA, and the commission was generally called the Paine Commission. It was supposed to articulate the national goals in space more carefully than had been done heretofore.

We studied the issues for two years, visiting most of the NASA research centers, meeting the scientists and astronauts. Soon we realized that the real challenge was the purpose of the space station that the US was going to build. After long debates and analysis, it turned out that the space station had only one real useful purpose, to serve as a place to assemble the vehicles for a manned round trip to Mars. The national plan then was to get to Mars by 2015, twenty years later.

The probable cost, and the later *Challenger* disaster quickly cooled general enthusiasm, and the former Soviet space station became the backbone of the "International" space station, which is now orbiting the earth, with no manned Mars mission in sight. Since we live on a planet that is rarely but regularly subjected to devastating impact with meteorites, the utility of a "life boat" for mankind makes some sense. But everyone thinks "not yet."

A Month in China

In 1985 I was able to fulfill a great wish of mine, to travel in China, to get to know China better. I had known parts of the Chinese world in Hong Kong, Taiwan, and other places, from various trips, but never on the mainland.

Several developments initiated a memorable trip, with lots to do, to learn, and to think about. At the time, I was at ITT, a corporate VP for research and technology. ITT, through its Belgian company, BTM, was building a factory in Shanghai to build modern electronic telephone switches for the Chinese Telecom administration.

This was considered a top priority business for us. Therefore, success was very important and the product was very high tech. So it was logical for me to see how it was going, make sure there were no unexpected major fundamental glitches, and what we should do to make success assured.

Requests were sent to BTM for my visit to the Shanghai factory. They agreed to the trip and got to work with Chinese authorities. When the Chinese realized that I was on the US Space Commission, they indicated that they would also very much like to talk to me about the Space Commission. As all this developed, Fred Ikle, US Undersecretary of Defense for Policy, developed the notion that I should give a talk to the Chinese military about SDI, the Space Defense Initiative, derisively called Star Wars. The Chinese indicated interest.

So the trip had a lot of pull as well as push. In the end, I spent a month visiting industry, universities, government departments, telecom, and microelectronics labs and factories. I traveled a lot, had many visits, met many people, including the cream of the technical community. I also learned a lot about China.

This was when Deng Jiaoping was remodeling China's economy and so its society. Officially, the Chinese were interested in showing their advanced ideas and their progress, and were generally friendly.

They wanted me to see the varieties of land and people and development, and made little attempt to hide their problems. They had clearly several objectives: First, to get me to help get a Chinese astronaut into the US space program (I promised to make Chinese wishes know in Washington, but made no promise of success – privately I thought it was unlikely). Second, they wanted to complain about US controls of export of technology to China. Most meetings started with the Chinese officials raising this issue. I had been told to expect it, so I responded by saying that it was temporary, and not aimed at Chinese progress, but to help defend China against "the big bear in the North" (which they understood well.)

The discussions of SDI with the joint staff of the Chinese Peoples Liberation Army, (PLA) were different, they were relatively belligerent, and had an anti-SDI line identical to the Soviets'. But it was interesting to see how the message was received, and it made everybody think more about strategic issues. My position was that a heavy ballistic missile defense was infeasible, and that moderate systems had a stabilizing effect. But the meeting was mainly very professional, giving lots of insight into their thinking.

A banquet hosted by Vice Mayor Li of Shanghai (Jiang Zemin was then mayor, later he became Chinese president) led to interesting exchanges about policy, and showed a very high level of policy understanding at the top Chinese levels. By the way, Li attended my customary return invitation for a banquet; I was told this was very unusual.

Every banquet started with a speech by the host, and every speech began with some Chinese proverbs, such as: every journey starts with a single step, and, a journey well-begun is half-finished. And the host's speech is invariably followed by the principal guest's speech, where the Chinese proverbs I had learned came in handy. And banquets start on time and finish when the host gets up (promptly after two hours). And never, ever ask for seconds, it will be too much to eat and drink. And watch the drinking of toasts.

I had three or more interpreters in my team, one from the US Embassy, one from ITT, several from my Chinese hosts. At the banquets they sat across from each other in the middle of the table. They would start telling jokes in Chinese so I demanded that all the jokes be translated for all of us. They complained, but it became a source of fun and educational.

Highlights of the trip were: fresh vegetables had appeared in farmers markets all over China, Deng's policy at work. The enormous size of the country, and its diversity were very impressive.

Most significantly, I saw China awakening.

While I was at ITT working in New York, I was approached by Josh Lederberg, a Nobel Prize biologist, who was President of Rockefeller University, to join that University's Industrial Advisory Board. I participated energetically, and I just hope it was as useful and interesting for the university as it was for me. I saw some of the beginnings of the start of the neurosciences, and met many of the key scientists.

About 1984 to 1985 I began to get worried because the buildup of the ITT technology slowed down, largely a matter of finances, and a return on the investment that was deemed too slow. I was disappointed, but had no ill feelings about it. To go on with the earlier speed would have been miraculous.

I began to think about alternative work, I was trying to get into something where I would build my net worth more quickly. I went to work with an investment firm. It was a group called Aetna, Jacobs, and Ramo, and I worked with them for five years. I was involved with finding people with good ideas that we could invest in. It was very interesting, not unlike the work at ARPA and ITT. But actually, my first challenge was something quite different, I was asked to run a small company that did contract manufacturing, building very complex electronics systems on circuit boards, working from blue prints and specifications.

The company was losing a lot of money, and I was supposed to fix it. In six months it went from losing 50 percent of sales to break even. Quality control of processes was the problem.

I had an interesting five years.

While my investment job was very much full time, I got permission to do some consulting because I wanted to keep my hand closer to the world of defense. I did some consulting for pay and some not, some pro bono.

Consulting has been a significant part of my career in a number of ways. I started out when I went to the NBS as a consultant to the chief of the Heat Division for theoretical physics and physical chemistry, which I thought was the most euphonious title I ever had.

If you run anything of any size and complexity, you are very busy, and you have to revisit these peripheral things in the outside world because that is where some of the inventions are, that is where some of the difficult problems are, and you have to be grounded in the broader context. Consultants can be enormously helpful.

The great thing about consultants is they add very much to your background, but you do not have to take their advice. I occasionally remind consultants when they want to pressure me, "You are being very helpful but nowhere in our contract does it say I have to take your advice." They do not like to hear that of course, but that is all right. I have made enemies that way, some time unnecessarily, but that is another story.

Los Alamos Lab

I have always found consulting as a very useful source of outside ideas. You get to work with somebody else on their problem that has its own logic it opens doors you would never think of opening.

The most interesting and altogether effective consulting job, I did was with Los Alamos, from about 1992 to about 2002. It got started in an interesting way. I had known the directors of the DOE labs for a long time. I have known all the directors of Los Alamos except very recent ones and Oppenheimer. (I never met Oppenheimer; but I heard him lecture a number of times.)

And the way that consulting for the Labs came about was that when I was DDR&D in 1990, I had a series of about ten very hard questions put to me by the secretary and the deputy secretary of defense. They were largely technical questions.

They wanted answers in a handy capsule, something that told them what those things were about. One had to do with stealth and another had to do with nuclear weapons. After working on this with my staff for a few days, I decided my staff was not helpful enough technically to get authoritative answers.

So I asked each of the three nuclear weapons labs, Livermore, Los Alamos, and Sandia to help. I called the directors of each place and we talked it over and they volunteered for certain questions and we made a sort of roster of who is doing what. The bottom line was that we had pretty good answers that I was not ashamed of or not worried about. All of the labs came through pretty well.

But the remarkable thing was that Los Alamos did much better than the other two. The Los Alamos people were more responsive; the people they put on them were better people. The answers came back in the form we asked for. So overall, the experience working with Los Alamos was a delight on this thing.

Therefore, when I left the government in 1992, I thought, one of the things I would really like to do is work with one of those labs. And so, I called each of the directors and said, "Hey, I would like to work with you. Are you interested? And if so how do we proceed?" And the most enthusiastic response came from Los Alamos. Sig Hecker who was the new director of Los Alamos said, "Very easy, come give us a talk about what you think Los Alamos should do, kind of a business plan."

I said, that sounds wonderful, just what I would like to try. That was a time when it was not obvious that Los Alamos (or any other weapons lab) would stay in the nuclear weapons business. They all thought they might get out of it, some were glad they might get out of it, some were scared they might get driven out of it.

When time for that talk came I had a large audience. In the talk I made two points.

I said, "There were really only two things that matter, one, you got to do, and the other, you should do. You got to do the nuclear weapons, because you are the most all-around capable lab, you have the most complete facility for all of it, and you started it all. And nobody will feel good if you get out of it, even though I know some of you want to get out it for various reasons."

Second, I said, "I have looked at your programs. I know what you are doing, and the thing you are absolutely top-notch in the whole country is computer-based modeling and simulation. You are doing an enormous amount of that for nuclear weapon design and still are, but also tactical things – air traffic control, road traffic, scientific modeling – and so on." I said, "If I were in charge, I would make the nuclear weapons task first, and computer modeling and simulation number two. I would put a lot of other things way behind on the list."

First of all, the country decided to keep nuclear weapons and to keep Los Alamos in the number one post, and computer simulation got a boost. By that time Vic Reis, my successor as Director of DARPA and as DDR&E, was at DOE in charge of the nuclear weapons program, and there was a danger that the country would neglect nuclear weapons, which I think would have been a disaster at that time. But Reis and Hecker and others rescued that, and Los Alamos did become indeed the lead lab. On the modeling and simulation, they kind of waffled. They did not bite into the apple, but kind of just sort of licked it. I think they became for a time the biggest computer-based model and simulation place in the world, but they could have been a lot better. And for my work for Los Alamos, I tried to help with that.

While doing this consulting, I spent a happy afternoon with the Connection Machine folks in Cambridge, MA. They were making a connection machine, which was a DARPA-financed machine, in the '80s. And they hooked it up for me to learn a bit how it worked. And they set up a program in such a way the machine calculated the flow of a hot gas in the pipe around a corner. They set up a very powerful workstation to drive the connection machine, and had set that up to inject tiny little balls into the stream of hot gas and then trace these balls where they went in that hot stream. They arranged it that I could easily inject balls at various places, with various properties, just by using a cursor and clicking, and so on. It was wonderful to watch this huge machine working silently and downstream with these fascinating consequences, and just to play with that. And I was sure that if I would want to, I could really learn a lot about hydrodynamics and maybe even discover something new.

And that demonstration was worth many thousands of well-written pages. There is a living tradition in any field that probably dies out after a few years. Who knows what is being done right now? The people who are really in the field really know that because they visit each other; play with each other's stuff. And that is the real advanced knowledge of advanced science; that is how you get it. Not by reading what has been published a year or two ago and written two years before that.

One of the most interesting of these consulting jobs was at the time when Hong Kong was still an independent state; they had a very strong economy. They were the biggest introducer of information technology, telephony, fiber optics and cell phones. And they set up an international committee to get advice on how to manage that: what to do; what to pursue; what were good bets or what were not. Charlie Kao chaired the committee. He was earlier at ITT, chaired the Fiber Optics KTSG, now he was Vice-Chancellor (which means he ran it) of the Chinese University of Hong Kong, a distinguished school, with a great faculty and a beautiful campus.

We met two or three times a year in Hong Kong, for two, three days each. We had the meetings at a hotel, and they were enjoyable. I think I was the only American in the group, a lot of Brits, many from Asia. One fellow was the head of the London School of Economics – very interesting person – a bunch of heavies. One of their early interests was in personal digital assistants (PDAs) – something they thought they could make and market. So we suggested they have a workshop on that, which they did and invited us to the workshop. That was a fascinating experience. They had us sit in the front row, the "advisors." The meetings were in English, of course. I remember looking back to look at somebody who was asking a question, and I realized, first of all, the room was full, maybe 150 people. Secondly, they were all young, in their thirties, very involved, full of energy. Most of them were men, maybe two women, maybe three?

Interestingly, one of the women, from Taiwan, presented a talk about a software piece, (which was actually quite small, only 150 kilobytes) a small program that turned handwritten Chinese into digitized Chinese characters. It is apparently simpler than it would appear. Everybody uses it now, and it is helped by several things. The Chinese do all the handwritten stuff that is to be digitized on paper that has ruled squares of the right size so you can put a character in. And second, there are standard ways to write every character, in terms of which stroke first, which way does the stroke go, up, down, sideways, where does the pressure go, how does it end, the flat end or the thin end, and so on; you all control that with what you write with, and that is caught in part of the software.

She made a demo on her laptop, started writing, and she also started reciting. A Chinese fellow next to me says, "That is a poem every Chinese learns." So she was writing that. She showed her laptop screen on the large conference screen, and almost every symbol was recognized by the computer. Now and then she had to choose one of two or three symbols, the machine could not tell which it was and all she had to do was click, and the right one was always in the list. This went on for a while. Well, she said, "Well, you can see it works." Well, we said, "Yes, it works." I talked to her later because I was interested, I had thought that the digitization of Chinese characters would be a major roadblock, but it had been solved. Most people did not then realize that a) that was a challenge, and b) it had been solved. And it has had enormous implication for the spread of computing in China.

When the independence of Hong Kong disappeared they stopped the meetings.

What I have found and still find is that people who invent something often are not quite good enough at explaining why somebody else should be interested and should pay money for it. It is primarily, I think, a piece of art in shaping an idea so it becomes really obvious. All ideas, once they are clear become obvious, most people do not go into the study of known things, except historians and philosophers, but most scientists want to work on unknown things.

When I was working on technology investments, I found that rarely is an inventor of something the best seller of it or the best manager of the company that is set up to implement the idea. So this is part of investing strategy and not that dissimilar in the consulting business. You have to help people to explain their idea, understand it, test it, verify it, make sure it works the way they are advertising their claim. If the solutions are there, what is the impediment for them to become policy? If the information is all there, then how come policy lagged so far behind the solution?

Money, for one – there are lots of things everybody knows should be done, but you cannot afford it. Perspective is lacking, the context is not examined, the horizon is unclear. I'm a great horizon believer and hence, the value of the French phrase *a tour d'horizon* – a tour of the horizon – which is a way to talk about something. Tour of horizon means you look at everything around and try to make it roughly equally important, not just mountains and valleys, but it describes the context, it is fruit, it is something you can make fruit salad with.

I think I look for the context now totally instinctively. I believe I learned to do that instinctively a long time ago. That is why for many, many years, everybody says I talk too long, say too much. These people want compact niche answers, bumper sticker answers. I tend to add context. The lack of interest in context, I think, is a form of intellectual laziness. People just focus on the one small problem at a time. Whereas I have found that often, the answers are easier when going one level higher and taking a wider horizon, and the answer may become obvious. It is a bit like a chess game; if you could see ten moves ahead, you would have a different vision of the game than if you see only three moves ahead, and that is where expertise, experience and a certain frame of mind comes in.

I have also usually found that by going from the problem given to you into a bigger problem, may solve the smaller problem in various ways. One is, you may decide "Is it important, really, or after all, do you have to focus on something else?" It is a perfectly good answer, or you may realize that the question you really wanted an answer for is in a different part of the universe. It is not chemistry; it is psychology. The big point here is that when one asks a significant question, one must decide what to accept as an answer.

A final conclusion, one that vibrates through my life, is that I am and was a builder, that was always my preferred stance. I learned over and over again that building something, an organization, or an idea, or an educated and trained individual takes time, often years. And I was always in fear of the many destroyers in life, events and people whose inattention or malice could undo a decade of effort in the blink of an eye. It makes one wonder how civilization ever advanced.

CHAPTER SIX

TO THE PENTAGON AGAIN AND THEN TO THE WHITE HOUSE

Shannon Comes into my Life

In 1989 I became interested in going back to the Pentagon, as the next DDR&E. There was a new Administration in town, President George HW Bush came in with new policies and a new approach, and in the Senate, a group around Senators Nunn and Lieberman wanted to change the top management structure in Defense, including a re-vivified position for the DDR&E.

I had been invited to the Republican Party Convention in '88 in New Orleans and testified before the Issues Committee on defense matters. That was an interesting experience which I am very glad I had. It was my first and presumably last dip into that brand of grand politics. Another important event came upon me unexpectedly. About that time, I met Shannon about some company business for Aetna, Jacobs and Ramo. While doing due diligence for a possible acquisition I was working on then, it was suggested I get in touch with her and get her expert input on the matter. Things went well, but our personal relationship developed more rapidly. We had to "fire each other" and focused on our private matters. We were married in June 1990, and the time since then has been the happiest of my life. She became my life companion, through good times and difficult ones. She also became a source of joy and of strength and understanding.

After George HW Bush won the election in 1988, Senator Tower was nominated to be the secretary of defense and failed of nomination, (rather rare event). It was a complicated and sad story. Next, Secretary Cheney was picked, and easily sailed through the nomination process.

It turned out they had some trouble filling the senior jobs in the acquisition top management. Finally, Don Atwood was picked as Deputy Secretary. Atwood had been a senior executive at General Motors. The Division of GM he had run mostly made batteries and automobile parts, but also some important space programs. John Betti was picked as undersecretary for acquisition, he was one of the three or four people who had remade the Ford Motor Company in the '80s and had run their European operations, and thus had been instrumental in reshaping Ford. This meant the two senior technical people under the secretary were from the automobile industry, but had been very successful in their jobs. Because of their background, they knew relatively little about the defense business or the defense industry. Then they had some problems filling the DDR&E position because the job had drifted into a much less significant position.

So after a while I decided, yes, that would be a challenge with the new way it was being set-up by the Congress, where major moves were planned to upgrade the position to its former influence. It was made to sound a lot more interesting than it had been in recent years. The first few DDR&E incumbents, Harold Brown, Johnny Foster, Bill Perry (who had a different title, Undersecretary, but equivalent influence) and others, had very powerful and influential positions, but the job and later incumbents got to be less effective.

So I put my hat in the ring. Then there followed a lot of amusing, as well as annoying things having to do with the nomination process. I had to fill out endless forms. By coincidence, all this happened just when my DoD security clearances had to be renewed. And all these processes went on independently and in parallel. People who knew me personally began to ask me if I was in any trouble with the police because they were being questioned a lot about me, whether I was a trustworthy person and a loyal citizen, and so on. And one inquiry came all the way from the White House!

A funny side story developed: a problem emerged when they looked at my resume. The problem turned out to be my membership of the Cosmos Club in Washington, DC, I had been a member since 1960. The Cosmos Club had been, since its foundation before 1900, a "men only" organization, and very appropriately the new administration said, "Nobody who was a member for long in an organization like that need apply. We just would not hire people like that." Well, fortunately I could prove that for the last twenty years I had supported, with small but real amounts of money, a faction in the club (led by member Patrick Hayes) that had worked very hard to change the membership rules to admit women. And "they" said, "Okay, you are innocent. We will look at you."

I think the deep underlying fact that influenced my decision to go for the position was that at time there was a large group of experts, and colleagues who were really interested, such as Johnny Foster, and some like Senator Nunn and Senator Lieberman, and myself, had rather clear and specific ideas what needed to be done in the Defense Department to improve and reorganize the acquisition process and the generation of new weapon systems and of new technologies for the DoD. We spent a fair amount of time articulating that, some papers were written on this by several people, including myself, and published, and discussed.

Meeting Secretary Cheney

Before I advanced to the confirmation hearings I had a number of sessions with Mr. Cheney. He was already the secretary of defense and these discussions were very constructive. Then I met briefly with Atwood and did not actually meet Betti until after confirmation, which was not ideal, but we just had to live with it.

I learned on a couple of occasions in my life that it is very worthwhile when people ask you to do a difficult, high level job, to think about this: "What do I need to be able to do that job the way it is presented to me?" With Cheney, the list of things I needed – good people, the ability to pursue the real problems without too much bureaucratic interference – I got all these secondary things. The one thing I did not get – which was critical and the first negative signal – was direct access to Cheney. I was stopped by Atwood, who said, "No, you have to come to me first." And I knew that meant: "You do not get to see the boss whenever necessary." It also meant: "You cannot do things that go against the major management system that you are under, but that you are supposed to fix," and to do the things that I thought needed to be done. This was very different from the set-up I had at ITT where I had good access to the top leader when needed. It did not bode well, and was a cause of lots of problems later.

I had a very easy confirmation session with the Senate. They allowed me to put a number of cards on the table. The most important of which was that I was then convinced, and still am, that applying standard commercial inflation factors to DoD R&D budgets means an actually shrinking R&D budget. And so, I put that issue into my opening comments. Of course I was immediately asked why I would say that? And of course, I had a prepared answer. After giving some real life examples, I was asked: "But why are the costs increasing so fast?" And I said, "Well, in the old days we used to try and climb mountains like the Appalachians, and now we are going after Everest, and that is much harder and takes longer and costs more... And if you want the world's best military, you have to pay a premium for being best." (Everybody understands this for motor cars: if you want a Rolls-Royce, you have to pay a large premium. And if you believe that all things considered you need the best quality defense establishment, for example to have successful wars with only small number of casualties, it will cost you like a Rolls. It is the price of having the best.) Well, this persuaded some people. However, when I later tried to turn that into real money for real R&D, I mostly failed, because it did not persuade the "budget hawks" and "bean counters" who had surrogates for their idea of the right numbers for inflation, but had no realistic pricing for "best in the world." (And many still do not understand the difference.) Nothing came of this fundamental idea. And it is still an important problem, in a managed market like the Defense budget, it is poorly understood how to measure the value of improvements in products that increase

Figure 6.1. The CNO Executive Panel ca 1990. I had been a member since 1970, and was invited to attend most sessions after becoming DDR&E. Admiral Kelso, CNO, center; author, 4th from right; Andy Marshall, 1st from right; Albert Wohlstetter, 2nd from left; Tom Evans, 3rd from left; Ben Huberman, 4th from left; Richard Cooper, 6th from left; Frank Fernandez, 2nd right from Admiral Kelso; Kathy Sullivan, our astronaut, 5th from right. US Navy photo.

the price of the product. There is a Nobel Prize in Economics hiding here, but no-one seems to be digging for it.

"Objectives" and "To Do Lists"

I thought that I had the charter to do a lot of major things. My list of priorities had about ten or so items; a long list is always dangerous, but there were some very important items that could simply not be ignored. One was to improve the acquisition process by making it more realistic in terms of judgments about what is feasible and what is not; and what is really needed and what is not. These actually were rather deep questions, and they are still plaguing the DoD. Nominally these issues were supposed to be analyzed and decided by the Defense Acquisition Board (DAB), then chaired by Betti, and vice-chaired by the Vice-Chairman of the JCS, Admiral Dave Jeremiah, and

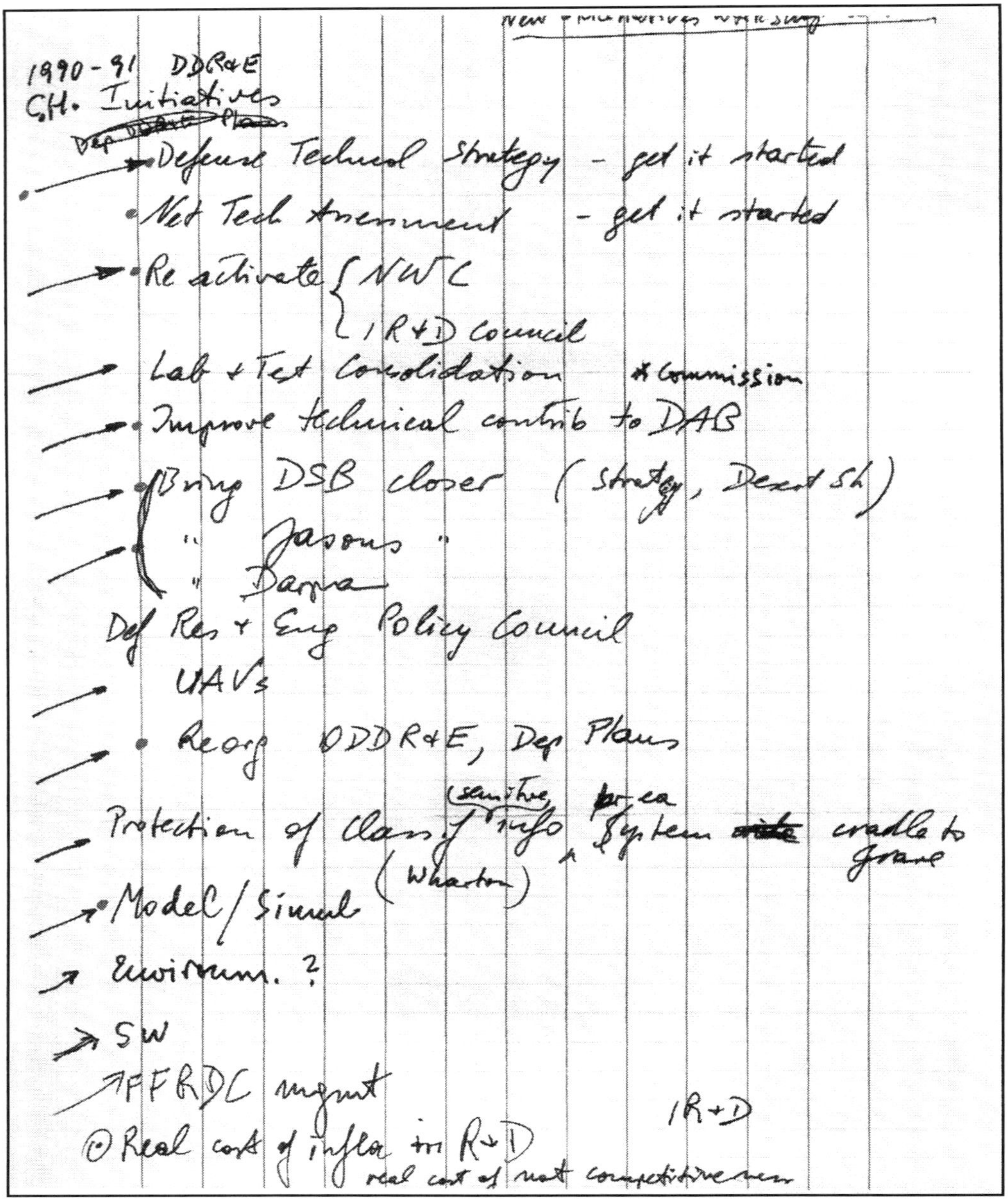

1990-91 DDR&E
C.H. Initiatives
• Defense Technical Strategy – get it started
• Net Tech Assessment – get it started
• Re activate { NWC
IR+D Council
• Lab + Test Consolidation – Commission
• Improve technical contrib to DAB
• Bring DSB closer (Strategy, Decad St)
" Jasons "
" Darpa
Def Res + Eng Policy Council
UAVs
• Reorg ODDR&E, Dep Plans
Protection of Classif (sensitive) info (Wharton), System cradle to grave
• Model/Simul
Environm. ?
SW
FFRDC mgmt
IR+D
Real cost of infla in R+D
real cost of not competitiveness

Figure 6.2. Our "to do" list, 1990. CH personal collection.

myself. Apparently, a powerful group. But in fact, it spent most of its time being project manager for the big programs, rather than to tackle the truly hard questions of "how good should defense systems be, and how many units do we need, and what should they cost." This problem dominated, and to some degree ruined, my tour as DDR&E, and the challenge question is still unanswerable, and its import persists. (As often in life, there is no strictly logical answer, but only a procedural and political one, and we were not well set up to search for these practical ways.)

To get clear about these large and complex issues, I planned a special off-site meeting to be held soon after my swearing in. It was finally held on April 26 and 27, very late, because Betti slowed it down as much as he could by requesting unreasonable procedures before approval. A clear example of "Process trumps substance." Eventually, it happened, but some of the initial momentum was lost.

The list of participants was a formidable one. It included John Foster, former DDR&E; Craig Fields, then Director of DARPA; Vic Reis, Deputy Director of DARPA; Harry Rowan, Assistant Secretary for Policy; Scooter Libby, also in the policy shop; Andy Marshall who ran the Net Assessment Office; Gordon Oehler from the CIA, a well known expert in armaments development; Bob Barker who was assistant to the secretary of defense for atomic matters (nuclear weapons); and several of my staff, as well as helpers and note takers. It went well, several attendees commenting at the end of the meeting: "this is breath of fresh air," and so it seemed to me at the time. The output was a "to do" list for my office and me.

My hope was that the results would be a road map for the carrying on of our business, and to some extent this is what it became. Yet, the basic problem that emerged quickly was that John Betti, the Undersecretary for Acquisition, had decided that his personal contribution would be to rewrite the regulations for acquisition, called DOD Directives 5000.1, 5000.2, 5000.3, and so on.

It was, and is, a morass of bureaucratic paper, necessary, but often an obstacle to manage innovation rather than a help.

At first, Betti and I had a moment of honeymoon when he said to me, "Charlie, I am process guy. I want to work on the process. You will be my substance guy, and you will tell me what is important." I thought that was great, except that it never really happened.

Nuclear Issues

But there were other very major things to get on top of. One function the DDR&E had (and it was reaffirmed by the new structure), was that he was Chairman of the Nuclear Weapons Council. That used to be so in the old days – in the '70s and early '80s – it was a very important and powerful committee, but it had atrophied over time. The Council was chaired by the DDR&E, and the other two members of the council were the Vice-Chairman of the JCS, who was responsible for the "needs" side; and the head of the Nuclear Weapons Program of the Department of Energy, which supplied the nuclear weapons and was the expert on feasibility. The job of that Council was to balance the requirements for nuclear weapons as expressed by the JCS and their system, with what was feasible to make, and when and for how much money as represented by the Department of Energy.

And the DDR&E was representing the secretary of defense in this work. That was the way civilian control of nuclear weapons was supposed to work, and when I worked there, it did.

Another grand priority in the acquisition domain that I pursued was to make weapon system more modular in design and construction. We in DOD, had great problems with upgrading existing airplanes and ships and tanks because we went at it without understanding the whole acquisition process adequately. Every tank was designed in a certain way and all of the pieces were designed for that particular tank model. Similarly for airplanes, and when we wanted to replace, for example, avionics electronics which needed to be done pretty regularly, it was very hard to do. It was difficult to change engine design in airplanes and we wanted to do that several times throughout the life of an airplane. (Commercial airplanes were designed to make this easier, and that system worked well, but the military at that time resisted this approach.) There was the same concern with ships, you want to change armament, guns and missiles, and you sure as hell want to change computers and electronics periodically. Making this easy would have brought a major transformation in how we do DoD acquisition.

Figure 6.3. Arriving at Applied Physics Lab, with my host, to speak about technology for submarines, upgrades and acquisition. Photo courtesy Applied Physics Lab.

However, no one was really very interested in making this change. The services were much too set on doing their own acquisition, and they were not interested in being serious about something abstract with long term impact. To actually enable a change as fundamental as this takes an iron will at the top of the department and that was not there, and from the DDR&E job you could not make it happen.

Another major challenge was to figure out how DARPA fits into DoD R&D. I had DARPA and the Defense Nuclear Agency reporting directly to me, and DARPA was mostly fun, but also grief, more about that later. The Defense Nuclear Agency was a serious operational group, mostly working on nuclear, chemical and biological weapons effects. The nuclear story was pretty clear, I was well informed about it, and it did not need a lot of my attention. Once the Iraq war started in 1990, the Defense Nuclear Agency became the key to working the chemical and biological weapons threats, because we expected the Iraqis to use chemical and biological warfare agents, and I stayed in close touch with their work.

To clarify the role of DARPA in DOD R&D, I was proposing that DARPA was mainly in the breakthrough business, and the services primarily in evolutionary development. Well, that sounds good, but the services did not like that, nor was it really that simple. The point was that DARPA had a process that very naturally led to high risk-high pay off efforts, which if successful would lead to major changes in how wars could be fought, while the Services, in fact devoted most of their effort and resources to gradual, across the board improvements of their forces.

In attempts to get the big picture articulated, I proposed the idea of a "trump card": That once in a long while, something technological happens, that is a like a trump card. It sweeps the table and nothing else matters. The best examples of this were the introduction of radar and of nuclear weapons, in World War II. These dominated the World War II military technology. (Of course there were also thousands of lesser advances wrought by each side's labs and industry, such as jet airplanes.) A new trump card which the US developed in the '60s and '70s was precision warfare coupled with the use of stealth aircraft. The war in Iraq clearly showed that this had happened. Indeed, this changed the tactical warfare game completely, a change that still has some ways to evolve.

In those days the DDR&E was required to have a substantial industrial background to be able to contribute a seasoned judgment of what was important and what was feasible into the acquisition process. I brought this background to the table, with eighteen years in ITT Defense and Aerospace, and as corporate VP for research and technology. But this personal contribution was kept suppressed by the processes governing the acquisition process. The key fact was that I could not affect the budget process very much. I could not switch money from program A to B except in DARPA and in DNA. This compares with the heyday of the DDR&E when Harold Brown and John Foster could block major weapons programs when these were deficient. The only substantive budget authority I had came from the Conventional Defense Initiative; God bless Senator Nunn who started the Conventional Defense Initiative, which gave the DDR&E $200 million every year to start new projects, and that was the only flexible money I had, and that was too small to leverage the DDR&E nominal budget of about $38 billion.

What was the flood of daily business like? Looking at my schedule from the 8th of March, 1990, just as I had been confirmed, I had fifteen things that I need to do urgently – near-term urgent. One was to get an office and to arrange the furniture; that was done. Then a congressional hearing of the Nuclear Weapons Council was coming up, and I worked with Bob Barker to get ready for that. Next, to start on a strategic plan for technology, that led to the very fruitful two-day meeting reported above, and a less fruitful written output – the strategic technology plan.

Then get organized for my first staff meeting which was on March 14, six days after I was confirmed. Then a congressional hearing on critical technology, which was to be on the 28th of March, twenty-five days later, and we got that done. Then Cheney came to visit to DARPA on March 15, which required preparation.

Then check out a candidate for my military assistant, and I was able to get Colonel John Klein of the Marine Corps, an experienced and distinguished individual, assigned to me. (He is now a member of Congress.)

Then get ready for a meeting of the Defense Acquisition Board (DAB). That was the key committee that managed the decision making of the acquisition process. The DAB, decided what to develop and buy, and how to do it. But, unfortunately the DAB had become the project manager for acquisition, and the deep questions of what and how much to buy became literally treated as distractions.

The Technology Strategy

A key assignment Betti gave me early on was to use my efforts and resources to produce a technology strategy for the defense department.

Little explanation was given when I asked what was wanted or what was to be done, and no discussion of this was allowed. There was no real understanding what the strategy was for, how it was to be used, or what it was to entail. Shortly after Betti asked me, he also asked the DSB to do the same. The DSB reported to me, but Betti did not explain to the DSB what he explained to me, namely that they were working for me on this, and were supposed to help with the effort.

So these two tasks became unrelated and competing efforts, pretty strange, for a supposed expert in management. One wonders whether it was malice or stupidity; life experience suggests that malice is really rare, so more likely it was the other. It came out pretty well in the end because we worked very hard to make it so. However, the perception through the Department was that it was a big fight, and one that I lost.

The deeper issue was what form such a strategy should take, and this was never addressed. It was a big problem, and the management was satisfied with slogans and bumper stickers. The assignment was so broad it could mean anything, so it meant nothing. On the other hand, some others wanted to make it so specific that it would have been a detailed step-by-step action plan rather than a strategy. In the end, a document was cobbled together that was not very good, I had no control over the final product. It was "Bureaucracy spinning wheels in uncontrolled gyrations, but progress is claimed."

And then, in June, disaster struck out of the blue. Craig Fields, the Director of DARPA, was fired while I was away on a brief scuba break.

Craig Fields was director when I became DDR&E, we had gotten to know each other before, and I was looking forward to working with him. But it turned into the most bitter story of my DDR&E experience. Betti fired Craig Fields without informing me, which I resented greatly. I had had several long prior discussions with Betti about Fields. Betti asked me what I thought about Fields, what kind of director he was and would become, and why he had so much public visibility. I told Betti that first of all, I thought he had the potential of being an outstanding director. I also said that to my mind the only issue was his public visibility, and that Fields and I had discussed this, that we were working on lowering his profile, and he was improving greatly and taking my guidance. So I thought the problem was fixed.

Fields is a brilliant and complicated man. He was doing very interesting things with the agency. But he also wanted to change US technology policy in several fundamental ways, and to get direct government funding for strengthening several industries. Now, every director of DARPA tests how much he, or she, can do (or else how much he/she can get away with.) Various directors found their limits in various places. I found my limits that had mostly to do with how to fight the conflict in Vietnam.

Craig found that he could not as director of DARPA change American industrial policy. So all of a sudden Craig was pushed out. When I got back from my break, I asked Col. Klein, my military assistant, what was new. He said, "Did you hear that Craig got fired?" I said, "No, what is going on?"

I found out that Craig had been informed a week before, while I was on vacation, that he was to vacate his desk at the end of the week I returned. And so I went to work on this problem. (I was reliably told that it was a very big political issue, but I got involved anyway. I believe in downward loyalty as well as upward loyalty, and nobody gets to fire anybody from under me without telling me first.)

So in fact he did not get fired, and I was able to assign him to be deputy DDR&E for planning. He could help with serious planning issues, and it would give him a little time to find what to do next, and perhaps to take some of the sting out of it. However, some people, I'm told, in the White House wanted to make an example of him and his ideas. So the Craig Fields experience was a bitter one, I could not stop it, I could not reverse it, only slow it down. It was a thoughtless move by the top management. I thought it was not only wrong from the point of view of supporting people working for you, it was also a bad idea politically, and it could produce a major political backlash.

But the real outside world intruded again: In August 1990, Iraq attacked Kuwait and occupied it. And we were involved in the war immediately. This opened up a variety of new challenges. DARPA quickly jumped in and Vic Reis, by then director of DARPA, did a number of clever and useful things, however these came too late because the war was over so quickly. One helped improve the armor of some systems very quickly, another improvised a simple method to identify friendly vehicles to our Air Force planes.

We also worked on deep-penetrator bombs, designed to destroy bunkers that are deep, deep down. I put the Air Force in touch with Los Alamos Lab. They had a great capability for this, using old gun barrels as the penetrating weapon, and that did very well, you really cannot stop it from going through the deep structures.

The problem I most enjoyed working on was related to the defense against Iraqi Scud missiles (cheap, inaccurate, missiles with large, conventional explosive warheads.) The Iraqis began launching Scud missiles against our troops and also against Israel. Various US Army systems were deployed to shoot them down, but had limited success. We started a study to see if one could shoot down Scud missiles while they are being launched, from airplanes with the right kind of anti-air missile. I knew that "of course you can." ARPA in the '60s did a lot of experiments on that problem and we knew how to do that, work which was originally motivated then by the Cuban Missile Crisis. As a start, the Air Force conducted a series of studies, simulations and experiments. It became clear that an Air Force tactic with two airplanes flying in a race track pattern so that one is always looking at a particular target was required. You have to know within some miles where the Scuds are. In the end the answer turned out that it could be done, but not enough of the right missiles were available at that time.

Federal Council for Science, Engineering and Technology

Meanwhile, I had a meeting coming up with the Federal Council for Science, Engineering and Technology (FCCST), which was made up of all the Government Department secretaries or their deputies, and was chaired by Allan Bromley, the President George HW Bush Science Adviser.

Bromley made FCCST into a very effective operation. At first, I thought my job would be to get Atwood ready for that meeting, but very early on, Atwood decided, "No, you, Charles Herzfeld, will represent me on the FCCST matters." So I was thrown in with all the department secretaries, Commerce, and Health, Education and Welfare, Interior, and so on. I represented Dick Cheney, the secretary of defense, and since I spoke for the most R&D money in the group and the biggest problems, that made some sense.

Later on, the major impact we had on Federal R&D was through the national high performance computing initiative (HPCI), one of the great ideas in the George HW Bush administration. (There were parallel initiatives in the Congress, including a significant one by then Senator Gore.)

Figure 6.4. Meeting of the Defense Policy Advisory Committee on Trade. Chair Norm Augustine, left; Deputy Secretary Atwood with back to camera; author, 4th from left; Asst. Sec. Macmillan, right. DOD photo.

Allan Bromley set up the national initiative with the National Committee on High Performance Computing, which became a panel of the FCCST. It was a big committee and a bit unwieldy, but it got a lot done.

DARPA led the technical arguments and had done a lot of homework, as had others. In some sense, it was a phase two or three for DARPA, because ARPA's ILIAC IV program of the late 1960s was the start of high performance supercomputing. Since I was the DoD representative on that, I spent a lot of time on it and found it was very interesting and rewarding.

The plan was that every department would put a pot of money into the program, but mostly DoD and DOE. And it would be a reasonably coherent program to do certain big things. The chairman of the high performance computing subcommittee was Eric Bloch, then director of the NSF and a former IBM computing manager, who had brought the revolutionary and highly successful "360-type" of computers to reality; I was the vice-chairman, as the DDR&E. Halfway through this project, Bloch left the government, and I became the chairman for an important period. I was able to set some of the objectives, and I was able to get the money from the OMB (Office of Management and Budget.)

We wanted a lot of incremental, which is new, money. We were spending about half a billion on high performance computing in all the departments, and we wanted to more than double it over five years, a big rise. First, I helped to craft the message. Then we went to the OMB and I was the lead arguer for the money. My argument was that every billion-dollar bill deserves at least one good technical idea. George Adams, the deputy director of the OMB (who had to approve the budget) said, "Now, that is a new idea. I like it, I like it a lot. I wish everybody would try that." So I said, "Well, we thought about it and it seemed sensible." And it was (and is) of course, you should spend big money only for big ideas; otherwise, you are limited to minor incremental moves. The new idea, one that was a DARPA contribution, was about "scalable computing." It described a design philosophy to create a computer in such a way that if you had a certain design, you could relatively easily make one ten times as a big, hundred times as big, or a thousand times as big. You could add the hardware, and the software, in such a way that you could put them together somewhat like Lego® blocks. People were hoping to be able to do that and in fact, that has happened to some degree. It turns out that it is not quite as powerful or efficient as one would like, but that was the next problem to work on. The first problem was to make it happen at all, and we did.

A Red Herring

Meanwhile, back in the Pentagon world, in December 1990 the notion was rumored that Cheney wanted to create a "Super DDR&E position." What that was supposed to be was left vague. Betti went on record that he could not accept this idea, and would rather resign. I was told by Cheney that I was in the running for the "Super DDR&E." At the Secretary's Christmas party, Cheney told me he "hoped I would get the

Figure 6.5. Reception for French Minister of Defense, 4th from left; Secretary Cheney, 3rd from left; author, 2nd from left. DOD photo.

Figure 6.6. Inspecting Seal Team West equipment in California. US Navy photo.

job," a job which was still to be defined. That was not a good sign. Later I realized that the main point was to get Betti to resign. Don Yockey, the principal deputy undersecretary, later became undersecretary. For me not much changed.

Not all the time was wasted on bureaucratic maneuverings. Let me give an example of a particular day, not really a typical day, but a good day, indicative of how interesting life occasionally was. Truly picked at random from a log of to-do calendars, the day was Monday, the 3rd of December of 1990. It started with the staff meeting with the secretary of defense at 9 o'clock, then a meeting with Mr. Kendall, (now the Undersecretary of Defense for Acquisition) an outstanding engineer who was one of my key deputies, on the status of stealth and counter-stealth, then at 10:30 to 11:30, a DAB pre-brief on a special program. Betti was in the chair, I was one of the two vice chairmen. The outcome of the meeting was inconclusive, nothing was decided. Then at 12:30, I had a meeting with Vice Admiral Reynolds of the Navy, on the lab consolidation, which was a hot topic, always on the agenda, but never really fixed: There were supposed to be fewer labs and hopefully better ones. At the end, we had fewer things that were called labs, but there were more buildings and more people. What we all tried to go for was really not working.

Then at 3:00 p.m., another meeting with the secretary of defense, which was classified and which was basically a replay of the DAB, but at the top DoD management level. All the senior people were there: Cheney, Atwood, the chairman of JCS, Assistant Secretary Wolfowitz, the Secretary of the Navy, the Secretary of the Air Force, the Chief of Naval Operations, and the Chief of Staff of the Air Force. It was a complicated issue, was supposed to last forty-five minutes, but ran over. (In my view the time was well spent, the issue was that important.) Then at 4:00, a meeting with Mr. Yockey, on the defense management review 922, which was also about the labs. Then at 5:00, a briefing for me on the pre-DAB on armor, anti-armor, the race between tank armor and anti-tank guns or rockets. And then I got to start (i.e., had time to work) on all of the papers I had to read and sign or write. It was a notable day, which I will not forget.

The peace dividend for ending the cold war, was part of the challenge. We were asked in the summer of 1990 to take a look at ten-year budgets and what big systems could be cut back, eliminated, or stretched. And we came out with savings for about ten years, at $300 billion a year, it was $3 trillion, with $1.5 trillion from acquisition, and we cut notionally about a trillion out of a potential ten-year budget. But life decreed otherwise, none of that happened then.

This analysis made troubles for the Iraq war of 1991, because people wanted to do away with systems that we needed during the war, and we had a little bit hard work to do to save those essential systems. Cashing in the peace dividend now was a mood change in the country. And it all became more bureaucratized under the guise of orderly management. Orderly management is always the enemy of inventive progress, in my view. "Orderly management" means no surprises, which means the enemy will surprise you. The way we used to run things we surprised the enemy a lot, with systems like stealth, precision warfare, and many others. However, in the end we were able to introduce some substantive sense into the proceedings.

Now this is a good place to tell about the one job I had when I was DDR&E that was really fun. By statute, the DDR&E was the technical conscience of all the special access (highly classified) programs, of which there were a large number. There was a two-star admiral Smith, who worked for the secretary of defense, who was in charge of managing these special access programs throughout the department. He started to keep me informed about the programs and involved with them. He liked what I was doing, so he was eager to get on my calendar and I usually was able to make time for him. The issues he brought me were always interesting, usually "good news" and exciting from a technology point of view.

How to Change the Bureaucracy

Managing the Defense Department bureaucracy for acquisition was a mess, because regulation had been piled on regulation; the system was unworkable. And as reported above, John Betti had decided to fix DoD Directive 5000 and the whole machinery of management that went with that. But the fault of this approach was that it focused on the mechanics and not on the content of the decisions. This effort was so bad and so wrongheaded that in the summer of 1990, four of us, the three assistant secretaries of the services for R&D and acquisition and I, met with Betti, and essentially forced him to have a private meeting with us.

I was elected the message bearer, a perilous role, but I did not mind, it was my job, and the message was, "John, you are missing the real process, how the big decisions are really made, with your working with DoD Directive 5000 and the paper process – the documents – that's not how it really works." And to explain how the exceptional process worked, I used a very important example that was a great success – the B-2, which was radical, completely different, and handled totally outside the regular acquisition system.

In this case, the Under Secretary for Acquisition who was Bill Perry and the Secretary of the Air Force and the Chief of Staff of the Air Force went to see the then secretary of defense and said, "Boss, we have a totally radical idea;" and the project was approved and proceeded to be very successful.

We tried to persuade Betti that this was a process, parallel to the conventional DoD 5000 process, it was how the really big decisions were made, and the rest you had to do to have professionally managed programs, thousands of programs, billions of dollars. Betti said, "No, a) I do not believe it, b) I'm going to change DoD 5000 and I'm going to make the whole department follow the DoD 5000." And I began to see that my quest in DoD would not be successful, no matter what I tried.

In the spring of 1991, Bromley came to talk to me and said, "Look, I would really like you to come work with me at the White House, on a very specific issue, set up something new to handle the growing problem of proliferation of weapons of mass destruction" And he described a dream job for me. I would report to him, the Science Adviser to the President and at the same time, to Brent Scowcroft who was the National Security Adviser to the President. This would be only the second time in the history of the republic that someone reports to the Science Advisor and the National Security Advisor. (The first was

Ben Huberman, in the Carter administration, a friend from the '70s. He gave me good advice for this new assignment.)

Bromley and I were quite friendly when I was DDR&E, and we spent some time working together. He invited me to come over and talk to his people at Office of Science and Technology Policy (OSTP) about R&D in Defense, and I made sure that he came and talked to the Pentagon people about national science policy.

To the White House via Russia

And I really liked that idea because, first, my job at DoD looked ever more like a dead-end, and I had many other things I wanted to explore and get my hands on that were multisided. (Most things in real life have this feature!) Besides, I knew Scowcroft well and he had already agreed to the move. I had worked with him in the Ford administration quite a lot. So I said to Bromley, "That sounds great." And Bromley said he would try and get me a deputy ship reporting to him. I told him I did not think that was likely, because it was a highly political issue and the number of deputies was set by Congress and to change that would be a horrendous thing, but he said he would try it. I said, "Good."

But he also said, "In May, I'm going to the Soviet Union. I'm taking a strong delegation and I'd like you to come along." And I said, "That sounds really marvelous and I would really like to do that." I had wanted to go to the Soviet Union for several decades, but it had not been safe to go, except as a part of an official delegation.

When I talked to Deputy Secretary Don Atwood about moving to the White House, I told him about the Russia trip and he said, "Good idea to move to the White House, you'll be able to contribute there. But you better move before you take the trip because I plan to be the first DoD person in the Soviet Union. I cannot go earlier, so you have to be gone by then." I said, "I understand that." That drove the timing a little harder than I wanted, but the trip was worth it. When the trip happened, Bromley took with him the deputy director of the National Science Foundation, one of Bromley's senior deputies, several other prominent technical people, and me. He had a whole bunch of things to negotiate with the Soviet government and agreements to ratify

Figure 6.7. Bromley in our little "palace". Photo by CH.

which had already to be negotiated, about science collaboration, and he wanted to take experts along.

So we were a group of about ten or fifteen people; several wives also went along. It was a fascinating trip. It lasted ten or twelve days, it was a very long trip. We flew to Moscow, I think we flew commercial. They put us all up in a very elegant, large "guest house" that was in the Lenin Hills area of town across the Volga from the Kremlin. It was a large complex where they had several very elegant guest houses for

Figure 6.8. Conference in the Kremlin.
Photo by delegation member, Robert White.

important Communist party delegation and diplomatic delegation visits. And we stayed in one of these – very nice rooms, lots of servants around who were all secret police, most likely.

We could not leave that place. I mean, it was nothing like house arrest, but we physically could not leave because the gate was locked and we were told, "Do not go outside, please." Guards were at the gates, and outside the walls, infrared detectors inside the walls that could cover every aspect of the wall so nobody could jump into the compound or climb out. It was a fortress. And they fed us quite well. It was interesting because we ate much better than the average medium level Russian functionary. There were a number of Russians who ate with us, and they were always trying to get left-over fruit because they had not seen apples that good. They were not great by American standards; but by Russian standards, they were marvelous – fresh fruit.

We ate well. We had good wine. Plenty of vodka if we wanted it, and we had very busy schedule, up early – meetings, meetings, meetings, and dinners in the evening; we were driven around to everything. We had several meetings inside the Kremlin, in these fabulous buildings in the Kremlin, the tsarist and communist palaces, and they are still in use.

The meetings were mostly signings of scientific agreements to share information. There were a number of new treaties between the US and Soviet Russia for joint technical work. Part was the beginning of an attempt to keep Soviet nuclear bomb people from working for anybody else. We tried to limit them to do things that we had some knowledge of. That time, we were worried that they would go to work for the Chinese, for the Iranians, for the extreme Muslims or others, and we really tried to do things to avoid that.

We took several trips to places that were interesting near Moscow, such as a huge and beautiful monastery. But the great insight of the trip came in the countryside, when I saw many carts with four wheels with rubber tires but drawn by donkeys! And I suddenly saw that Russia was still a third world country, more like India than like Germany.

We took a trip to Baikonur where the Soviet missile launch center is located. They invited us to a launch of a manned space flight, which was very interesting. The public buildings were very much more modest and smaller than our NASA manned launch areas are.

The space launch was very interesting and the remarkable thing was it took place exactly at the time that they had announced, no hold, no postponement, and it went flawlessly, like a bus leaving. This was a time when our space launches were very often delayed for technical reasons. That made a big impression on me. They were also very proud of their first manned space flights that Yuri Gagarin and others made. He was dead by the time of my visit, but his pictures were everywhere and there were medals engraved with his faced. I received one of these. Nothing very fancy. They and the Chinese really give cheap memorabilia – probably 20 cents worth, made of tin. But that is all right, they are worth keeping.

In Baikonur we went out on the missile launch range to look at civilian rocket launches. Remarkably, the Russians allowed visitors near the launch pad while rockets were being fuelled. We also took a tour of the facility. Somebody was assigned to go with me who was clearly a senior intelligence guy. I had seen lots of pictures of that place, and at one stage I said, "I wish we could go here to the left road rather than to the right." This guy laughed and said, "Well, not today, we are not." This was because all the military ballistic missile stuff was along the left road, which I knew, and which he knew that I knew. So this was a joke among "colleagues." And he said, "I'm sure you have seen plenty of it." I said, "Yeah, I have not had enough – not close enough." He said, "Yeah, yeah."

Anyway, after Baikonur we went to Akademgorodok, the academic city in Siberia. We spent several days there. They showed us around several scientific institutes, which was really fascinating. The people there were very smart scientists, but they had not been doing much for some time. The papers they presented to us were old and the experts knew that there was not anything newer available there. Russian sciences had ground to a halt by that year, 1991.

A vague sense of impending disaster was in the air, and in fact the Soviet Union dissolved a few months later. There were lots of receptions at which every one of us had a chance to talk to some high level Russians about various things. It was clear that opportunities were made to talk to us privately.

One feature in those days was that the Soviet labs had no modern computers in them at all. Part of the US policy in the Reagan administration was not to sell them any computers – and it obviously worked. Of course, we did not see the military labs, where there were some computers, but only old ones by our standards.

But the civilian labs had none. The academicians, however, the senior people who ran the labs, had modern PCs. One asked me, "I have a (Intel) 386 machine, is it worth getting a 486 machine?" for his own personal computer that they got for him. I said, "Well, it depends what you want to do with it. What do you want to do?" I had just graduated to the 486 machine. So we had interesting discussions and he did not really say what he wanted to do with it, but it was clear he could have gotten a more modern machine if he wanted one.

We talked about science. We talked about the political evolution going on in the Soviet Union; and the more senior the Soviet people were that we talked to, the more pessimistic they were about the continuation of the Soviet Union. A number of these told us, "We do not know if we are going to make it. We think we may not." Most of the senior Soviet people we met thought they had lost the Cold War, while most of the US folks I knew then were not sure that we had won it. The Soviets had major food shortages, major civil unrest, political problems, economic problems, trade problems. They were very impressed by our military.

When we got back to Washington, I settled into working in the White House. (Not physically, our offices were in the Old Executive Office building next to the West Wing of the White House, but we regularly went to meetings in the West Wing. I think Bromley was one of the most effective science advisers the country ever had. He got to see the president whenever he really needed to. President George HW Bush was great. He came to every meeting of his PCAST, the President's Council for Science and Technology that I was working with. The President regularly came with a notebook, took notes and he asked questions and wrote the answers down. The next time he would say to some member, "Oh, by the way, doctor, excuse me. Last month you said such and such. Can you tell me more about it? I'm really interested." We had our meetings in the Roosevelt Room in the White House, which is a wonderful room – at that time it was the Theodore Roosevelt room, by the way – with lots of Western art and it is just a few feet away from the Oval Office. The President would walk in and say, "I'm sorry I'm late. I was on the phone with somebody." I got my assignment done, and soon after I left, a new system of non-proliferation activities began.

Bromley had a wonderful style to do these things and the people he had on the committee were excellent. Working at the White House is different in lots of ways, you know that you are working for the person who runs the whole government, the head of our state.

A little example: in the Pentagon, they have a daily sheet that comes out with all the new stories about the Pentagon, it was called "the early bird," it was about half an inch thick. In the White House, they had the same kind of press summary, but for the whole government – it was two inches thick – every day. The PR-related people always scanned it immediately and some of the senior people got it delivered at home before they come to work.

Besides the important news stories, it had in it such items as the jokes about politics, the President, the government, things that were said last night on TV. What did Letterman say; what did Jay Leno say? It was interesting, because it showed that the White House is a place where you worry about the whole country – everything, all the time, all major crime, all big fires, earthquakes – everything. (I also learned that the late night TV comics are significant indicators of the public mood.) Some gruesome stories about crime or accidents were also in it, or if something really bad happened somewhere. Nobody reads all of it, but many people read some of the pieces. It is all part of our system of government.

Finally, in February 1992, it was time to go, to leave government. Another major change in my life was about to take place. Then the consulting jobs came, and I was busy again. But it had been a good time, a tough time, a time of service to the country that I am proud about, and that I will never forget.

Figure E.1. Relaxing in Bonaire. CH personal collection.

Endpapers

Coming to the end of the story, makes again an opportunity to reflect. Much has happened and many people have helped. The perspective gained through the writing was important; I sleep better now than when I started, and I am more at peace with myself, other people and my life. A stroke in April 2010 slowed me down a lot, but I am able to manage daily life, and personal and professional aims and obligations. As I have said for most of my life "Life is great, I would not have missed it for anything." And I say so still. I had to stop scuba diving, but am relearning my snorkel skills, and I still love the oceans.

I am proud of most things I did, though not all, and I am proud of the motivations, and mostly of the results. And I am astonished, and just a bit scared about the consequences of starting "to tie all the stuff together" when we started the ARPANET.

In 1965, we hired Bob Taylor as new director for IPTO with the explicit goal of trying seriously to build a network of computers and associated communication devices, using the technologies of time-sharing, graphics and communications that had been developed by IPTO before. No one had tried this on a large enough scale, using packet switching. It was far from obvious that it could be done, and many voices weighed in with major opposition and doubt. However, the idea was attractive and plausible, and we freed up new resources for this. And if it turned out to be too hard, I preferred that we should learn this sooner rather than later.

So we proceeded with this. The details have been well told and do not need repeating. But the project was a success. ARPANET worked, and showed it could be done.

Figure E.2. Author receives from President Clinton, for Bob Taylor, the National Medal of Technology, for the start of ARPA net. March 14, 1999, at the White House. DOCommerce photo.

After I left ARPA, the people there continued the stream of work and, later built the Internet, based on the critically important Internet protocols, "IP" and "TCP", developed by Bob Kahn and Vince Cerf, both at ARPA after my time: in my judgement, the most potent engineering invention of the 20th century. And that was a project that, over time, has changed the world.

Did I expect what has happened? Of course not. Did I have great hopes? Yes, but on a more limited scale; I thought this work might revolutionize the production of books, as well as the structure of scholarly writing and publishing, I saw more clearly than most the need to organize non-fiction books differently, but I did not foresee the change in production and distribution of books. Nor did I expect the revolution in work and play that was enabled by the Internet and Web.

When I was a boy, I wished for a library with all the books in the world, and a magic carpet that would take me anywhere, clearly inspired by

Figure E.3 Author with Bob and Patrice Kahn. Photo by Susanne Ollmann.

the classic Arabic tales. And to an astonishing degree, modern science and engineering is making these dreams possible.

Most of my life I have prepared for wars I thought would be unavoidable, and worked hard to prevent war fought for unworthy causes.

It seems to me now that several lessons stand out clearly: Real problems are usually quite complicated, and are not easy to solve. Balanced judgments are usually a good idea. Big changes take a very long time. Actions just for appearances are an unworthy goal, but are so easy to settle for that very many people are tempted this way. Most important problems are a mixture of science and technology, of policy choices to be made, and political decisions to be made. Over time, it has become clear to me that usually, politics will trump policy and both will trump science and technology. And I think that is just right for a society steered by the will of the people, which is expressed though politics.

Figure E.4. Shannon toasting us at my 80th birthday, 2005. Paul Kozemchak, an old friend, in center; and Craig Fields, at extreme right. Photo by Susanne Ollmann.

Importantly, I am delighted with the new friends I made through the process of writing this book; that alone was worth the effort. I appreciate all my friends more than before, and my efforts to name as many people as possible who helped throughout my life, made me appreciate the extraordinary number of wonderful people who contributed to all we tried to do.

Charles M Herzfeld
Silver Spring MD
January, 2014

INDEX

36657919R00164

Made in the USA
Lexington, KY
29 October 2014